How To Become A Professional Engineer

D. G. Sunar, Ph.D.

PROFESSIONAL PUBLICATIONS, INC.
SAN CARLOS, CA 94070

In the *ENGINEERING REVIEW MANUAL SERIES*

Engineer-In-Training Review Manual
Quick Reference Cards for the E-I-T Exam
Mini-Exams for the E-I-T Exam
Civil Engineering Reference Manual
Civil Engineering Quick Reference Cards
Seismic Design for the Civil P.E. Exam
Timber Design for the Civil P.E. Exam
Structural Engineering Practice Problem Manual
Mechanical Engineering Review Manual
Mechanical Engineering Quick Reference Cards
Electrical Engineering Review Manual
Chemical Engineering Review Manual
Chemical Engineering Practice Exam Set
Land Surveyor Reference Manual
Expanded Interest Tables
Engineering Law, Design Liability, and Professional Ethics

In the *ENGINEERING CAREER ADVANCEMENT SERIES*

How to Become a Professional Engineer
The Expert Witness Handbook -A Guide for Engineers
Getting Started as a Consulting Engineer
Intellectual Property Protection -A Guide for Engineers

Distributed by: Professional Publications, Inc.
Post Office Box 199
Department 77
San Carlos, CA 94070
(415) 593-9119

How to Become a Professional Engineer

Printed in the United States of America

ISBN: 0-932276-52-0

Current printing of this edition (last number) 6 5 4 3

PREFACE

This book is written for the engineer or engineering student who wishes to learn how to become registered as a professional engineer. It will serve as your companion and guide through the entire registration process, from application to receipt of your license.

Beginning with a brief history of registration, it shows you the benefits of registration from both the professional and the personal points of view. It then takes you through the registration process, with practical suggestions for success at every step.

The requirements for registration, in terms of education, experience, and examination are discussed. You will be shown how to apply for registration, how to document your experience, and how to prepare for the examinations.

Each of the examinations (EIT and PE) is discussed in detail. You will learn how to increase your chances of passing with thorough preparation, control of anxiety, and point-maximizing strategies for both examinations.

D. G. Sunar, Ph.D.
Belmont, California
May, 1985

ACKNOWLEDGMENTS

In writing this book I have been helped by a number of people, and I would like to take this opportunity to thank them here. Michael R. Lindeburg, P.E., can be credited with many of the ideas, and he supplied me with information and support throughout the writing process. Rhonda A. Jones has been especially helpful with all technical details, including graphics. Sandra Park made useful comments and corrections. I am grateful to all of them.

I would like to acknowledge here the pioneering work of John D. Constance, who has devoted a great deal of effort over the years to helping engineers along the road to registration. I am grateful to him and to all the other authors whose work I consulted in preparing this book.

D.G.S.

TABLE OF CONTENTS

As its title suggests, this book is a "how to" book, designed to give you full information about how to become a registered professional engineer. In the following pages you will find answers to many of your questions about the process of registration, as well as many helpful suggestions for applying and preparing for the examinations.

REASONS FOR REGISTRATION

The first question you may ask about registration is, "Why should I register?" Particularly if you are a recent graduate of a rigorous engineering curriculum, you may feel that your degree speaks for itself and that further testing and certification are superfluous. However, the issue is a good deal more complex than that.

A Brief History of Registration. Ever since the Code of Hammurabi, civilized societies have held individuals responsible, not only for their actions, but for their products. Under that ancient code, a craftsman whose product caused a user injury was liable to have the same injury inflicted upon him in return. Over the centuries, the emphasis has shifted from punishment of the guilty party to compensation for the victim, but the same principle is involved in the liability laws that are in effect today.

No society can leave questions of responsibility and compensation (which are ultimately questions of justice) to the individuals involved. As each party to a dispute seeks support from others, the conflict may spread, so that the stability of society itself is potentially threatened. Thus all societies have legal and judicial systems of some sort for the resolution of individual disputes, and all societies regulate the performance of certain kinds of roles and services.

As societies and their division of labor have become more complex, more and more specialties and professions have developed. One hallmark of practitioners of most of these specialties and professions, from barbers to medical doctors, is that they offer services to the public based on specialized training, skill, or knowledge. The question then arises: how do we know that a particular individual is a competent barber, doctor, or engineer?

In the early Middle Ages, King Roger of Normandy ruled that a man could not practice medicine unless he had been found competent by a group of practicing doctors. Even in those early times, when medical knowledge was scant (and often wrong) by today's standards, doctors were perceived as having life-and-death powers over their patients. King Roger attempted to protect the well-being of his subjects by seeing to it that those who practiced medicine had the necessary knowledge and skill, and he left it to practicing doctors to judge the competence of aspiring doctors. This was one of the earliest precedents for the principle of professional peer review, which is still applied in the present day.

The Middle Ages saw a rapid spread of the idea of regulating the practice of craftsmen and professionals. Most trades and crafts became organized through the guild system. In this system, one could become recognized as a master of one's craft (for example, a master builder) only after a long period of apprenticeship, followed by a period of relatively greater independence and responsibility as a journeyman. Master status was conferred by the older masters through a process of peer review.

Conditions have changed greatly, particularly through the introduction of formal education to replace childhood apprenticeship. However, we still see the journeyman phase of training in the trades, where it retains its old name. We see it in the professions as well, where it is usually called internship.

It was not until the nineteenth century that governments began to get involved in the certification of experts of various kinds. As universalistic ideas of justice and new concepts of the functions of government began to spread, the need for protection of the public from fraud and incompetence began to be felt.

Like all other types of professional registration, registration for engineers was motivated by concern for the public welfare. In the United States, it began in turn-of-the-century Wyoming. There, maps and plans of homesteads, streams, canals, and waterworks were being prepared by practically everyone other than engineers and land surveyors. For instance, lawyers, real estate brokers, notaries, insurance agents, pawnbrokers, and many others prepared and submitted maps and plans. As one might expect under these circumstances, many of the maps and plans were grossly inaccurate (some of them purposely so), leading to many property losses and other problems.

The State Engineer, Mr. Clarence T. Johnston, set out to prevent these abuses. In a pioneering effort, he obtained cooperation in introducing legislation which would require all engineers and land surveyors to register with the state. Despite the opposition of various groups, the bill passed in 1907, marking the beginning of professional registration for engineers in the United States.

Abuses of land and waterworks surveys were not confined to Wyoming, of course. Other states, beginning with Louisiana, enacted laws requiring the registration of engineers and land surveyors. Awareness of the public responsibility of engineers as designers of buildings, roads, bridges, and industrial and other equipment of all kinds began to grow. Today, all fifty states and four other jurisdictions (the District of Columbia, Puerto Rico, Guam, and the Virgin Islands) have regulations governing the registration and activities of engineers.

There has been no attempt to create a national system of registration. The authority to regulate the practice of professions in the interest of public welfare is part of the police power which is reserved to the states under the Tenth Amendment to the Constitution. Each state, having a different history and a different set of problems, enacted legislation designed to meet its own needs. The result was, in the beginning, a widely varying set of requirements and procedures for registration.

Gradually, the states and jurisdictions recognized the need for greater uniformity. Today, most states employ essentially the same testing procedures. This uniformity is largely the result of the efforts of the National Council of Engineering Examiners (NCEE), a body which is composed of representatives of all the state and jurisdiction boards of examiners.

Regardless of the degree of uniformity among the various jurisdictions, all have followed the venerable principle of peer review. Although it is carried out under the control of the state government, the review (by means of examination and evaluation of professional experience) is always carried out by competent members of the profession.

The Situation Today. With the excellent preparation for a career in engineering that is being offered in many colleges, universities, and engineering institutes today, you may wonder whether the issue of protection of the public is still a vital one. Leaving aside the question of fraudulent practice, aren't the engineers of today competent enough to function without state supervision? A few quotes from actual examination interviews may be instructive.

One applicant, applying for registration on the strength of his many years of "engineering" experience with a branch of the U.S. Armed Forces, faltered on the most basic types of questions.

Interviewer: What is a horsepower?
Applicant: 33,000 pounds.
Interviewer: Do you mean that a horsepower is the same as 16-1/2 tons?
Applicant: No, sir. It is twice as much.

A second candidate, listing experience as a heating and ventilation engineer, had similar problems.

Interviewer: How much heat does it take to make a pound of steam?
Applicant: It takes 33,000 BTU per pound of pressure per minute of heat transfer.

While these two candidates may seem amusing, a third reminds us forcefully that ignorance may be a public menace. This man represented himself as the designer of large roller coasters for fairs, expositions, and amusement parks.

Interviewer: How do you calculate the stress in the structure?
Applicant: I know from experience how large to make all the pieces.
Interviewer: What calculations do you make?
Applicant: I have a formula in my office for the force of a car going over a curve.
Interviewer: Do you employ any other formulas?
Applicant: No, sir. That is the only one I need.
Interviewer: What safety factor do you use?

Applicant: About 90%.
Interviewer: Do you mean your safety factor is less than unity?
Applicant: Yes, sir, less than unity.
Interviewer: Why do you do that?
Applicant: To give the customers a bigger thrill. They have to hold on when they go over a curve. When a car is going fast, you don't need 100% safety. You get across before anything happens.

These individuals looked upon themselves as engineers, and they felt qualified to take the registration examinations. If we remember that engineers are involved in the design and construction of almost all the buildings, roads, bridges, conveyances, and equipment that people use in the course of their everyday lives, it is evident that the public continues to need protection from poorly trained persons who do not know their own limitations.

Professional Reasons for Registration. It is not only the general public which benefits from the registration of engineers. The engineering profession itself benefits in many ways from the registration system.

One of the most important benefits is that the title of "professional engineer," or any of the other registered titles, is protected from unauthorized use. A person who has invested many years into the education and working experience required to be a professional engineer is rightfully concerned about protecting that investment. If individuals who have not acquired the requisite education and experience were free to call themselves professional engineers, the public would have no reliable way to distinguish between truly qualified engineers and those who were simply using the title. The registration process serves to recognize and protect qualified engineers from inappropriate use of the title by unqualified persons.

Not only the title, but the actual practice of engineering is protected by registration laws. This is the heart of the law, from the point of view of protection of the public. It is also essential to the engineering profession, in protecting licensed practitioners from unfair and unqualified competition. With legal recognition and protection, the profession can take steps to prevent unlicensed individuals from advertising themselves as engineers, and otherwise to restrict their actions and practices.

Regulation of the practice of engineering also prevents encroachment by other professions. In the past, professional interest groups have attempted through legislation to reserve certain areas and activities to themselves. For instance, legislation has been proposed to reserve the sanitary field entirely to physicians; to prevent anyone but accountants from making financial reports; to allow only lawyers to prepare contracts and to engage in arbitration; and to reserve the appraisal of land and buildings to real estate brokers. These encroachments and restrictions on legitimate functions of engineers have been fended off largely through the legal regulation of engineering registration and practice.

Registration also helps to maintain the good reputation of the engineering profession in the public mind. A public which is protected against fraud and incompetence rewards the honest and competent practitioners of a difficult profession with the respect which they deserve.

Registration encourages the continual development of the technical competence of members of the engineering profession. One effect of registration laws has been an increasing focus on education as a crucial element of the engineer's qualifications. In earlier days, engineers usually learned their skills through experience, not formal education. Even today, in some states, there is some opportunity for a person with extensive experience to substitute years of experience for years of education in qualifying for registration. However, greater emphasis on education helps to assure more dependably adequate training for engineers.

As the importance of education has increased, a system of accreditation for engineering programs has been inaugurated. Accreditation has led in turn to a general improvement in the programs of instruction, the quality of teaching, and the equipment and facilities available to the engineering student. As a result, engineering students are now emerging from their educational programs better and more uniformly qualified than in the past.

As examinations have been made part of the registration procedure, experienced engineers have found it necessary to bring themselves up to date in their discipline in order to qualify for registration. This also helps to raise the general level of technical competence within the profession.

Regulation and registration have had a major impact on the development of professional consciousness among engineers. Perhaps because of the relatively late development of standardized training and education, engineers have been slower than members of some other professions (for example, medicine and law) to develop a professional identity and to acknowledge the importance of professional concerns. Through registration, an engineer becomes part of a readily recognizable group which is defined in terms of its standards of competence in a particular field.

This sense of identity forms a basis for an interesting paradox: as the profession has become more regulated by law, its internal mechanisms for self regulation and self discipline have also increased. Professional consciousness and identification with the professional group have led to an increasing concern with the maintenance of standards for qualification and practice.

Personal Reasons for Registration. What is in it for the individual engineer? Registration brings a number of benefits and potential benefits to the individual.

In most states, you cannot do business or advertise yourself as an engineer unless you are registered. If you are employed by a manufacturing or other type of non-service firm, you may be exempt from this rule, in the sense that you may do engineering work. However, you cannot function as an engineer outside the firm. Therefore, if you wish to supplement your job with outside consulting, you must be registered. If you are not, your work will not have legal authority, and you may have difficulty collecting your fees.

If you wish to join a consulting firm as a partner, or perhaps establish your own consulting or engineering firm, it is legally imperative for you to be registered. Registration also provides the important side benefit of allowing you to participate fully in professional engineering societies, such as the National Society of Professional Engineers (NSPE). Membership in these societies is very helpful in terms of referrals, advertising, and other forms of business assistance.

Even if you are not interested in branching out into independent consulting, you may find that registration brings many advantages. Many firms will not promote an unregistered engineer into senior engineering positions. When making bids for contracts, many firms believe it is to their advantage to list a professional engineer, rather than an engineer in an exempt status, as the person responsible for the project. Also important from the point of view of all employed engineers, many firms pay a higher salary to registered engineers in all grades.

If you are engaged in engineering work, but do not have a traditional engineering degree, registration may be especially important in establishing your professional identity as an engineer.

Registration can also provide a kind of long-term insurance. You may be content with your job at the moment, but an economic downturn or other situation could lead to the loss of your position. Under such circumstances, professional registration could make it easier for you to find a new job, or to work as a consultant while seeking a suitable position.

Likewise, if you wish to remain professionally active after retirement, registration would make it possible for you to join a consulting firm or to act as an independent consultant. These possibilities could be important from the financial point of view, as well as allowing you to remain productive after retirement.

WHO NEEDS TO REGISTER?

All engineers who wish to practice as consulting engineers must be registered in each state in which they wish to do business. There are no exceptions to this rule. All engineers who offer their services directly to the public must be registered.

Engineers who carry personal responsibility and authority for projects involving public safety (and most engineering projects potentially involve public safety) must be registered, unless they are working in an exempt status. If the engineer is exempt from the requirement of registration, another person or entity (for example, the employer or supervisor) must assume responsibility for evaluating the engineer's qualifications and work, and for the safety of the project.

Although specific rules differ somewhat from state to state, there are five general categories of engineers whose work is exempt from the licensing regulations. These are:

1. Engineers who work for the U.S. Government.
2. Engineers who work for a manufacturing corporation or a corporation engaged in interstate commerce. (This is the "manufacturer's exemption" or the "industrial exemption.")
3. Engineers who work for a public utility or public service corporation.
4. Engineers who work under the supervision of registered engineers.
5. Engineers who design or build structures or equipment for their own personal use.

Although these categories are currently exempt in most states, elimination of some exemptions has been proposed in various quarters. The movement toward consumer protection will undoubtedly give impetus to these proposals. At least one state has done away with the manufacturer's exemption, and it is anticipated that other states will reduce exemptions in the future.

Considering the advantages to be had from registration, and the uncertainty of the future of the exempt categories, it is advisable for every qualified engineer to become registered.

According to NCEE statistics, there are currently over one million people in the United States who can be called engineers. As of 1984, about 340,000 of these engineers were registered in at least one state. In addition, there were about 241,000 out-of-state registrations, indicating that a large proportion of registered engineers are active in more than one state. Engineering registrations have been on the rise since 1937 (the first year for which systematic nation-wide statistics are available). Figure 1 shows this upward trend in registration graphically.

THE ADMINISTRATION OF ENGINEERING REGISTRATION

Registration of engineers in each state and jurisdiction is administered by a board of engineering registration (the specific name of the board varies from state to state). The composition of the boards may vary according to state regulations, but each has some members who are professional engineers. In each state, the board is responsible for evaluating applications for registration, administering examinations, awarding licenses or registered status, and enforcing state laws regulating the professional practice of engineering. In some states, the board also acts in an advisory capacity to the legislature. As technology and public needs change, the board recommends appropriate changes in the registration system.

The structure and functions of the state board may be somewhat different in a state such as Delaware, where the board is not part of a state government agency.

The fifty-four boards have joined a voluntary association composed of representatives from each board. This association is called the National Council of Engineering Examiners (NCEE). NCEE has been a powerful proponent of uniform practice in the registration of engineers. Examination procedures in most of the states now follow NCEE recommendations, and almost all of the states use the standard examinations prepared by NCEE.

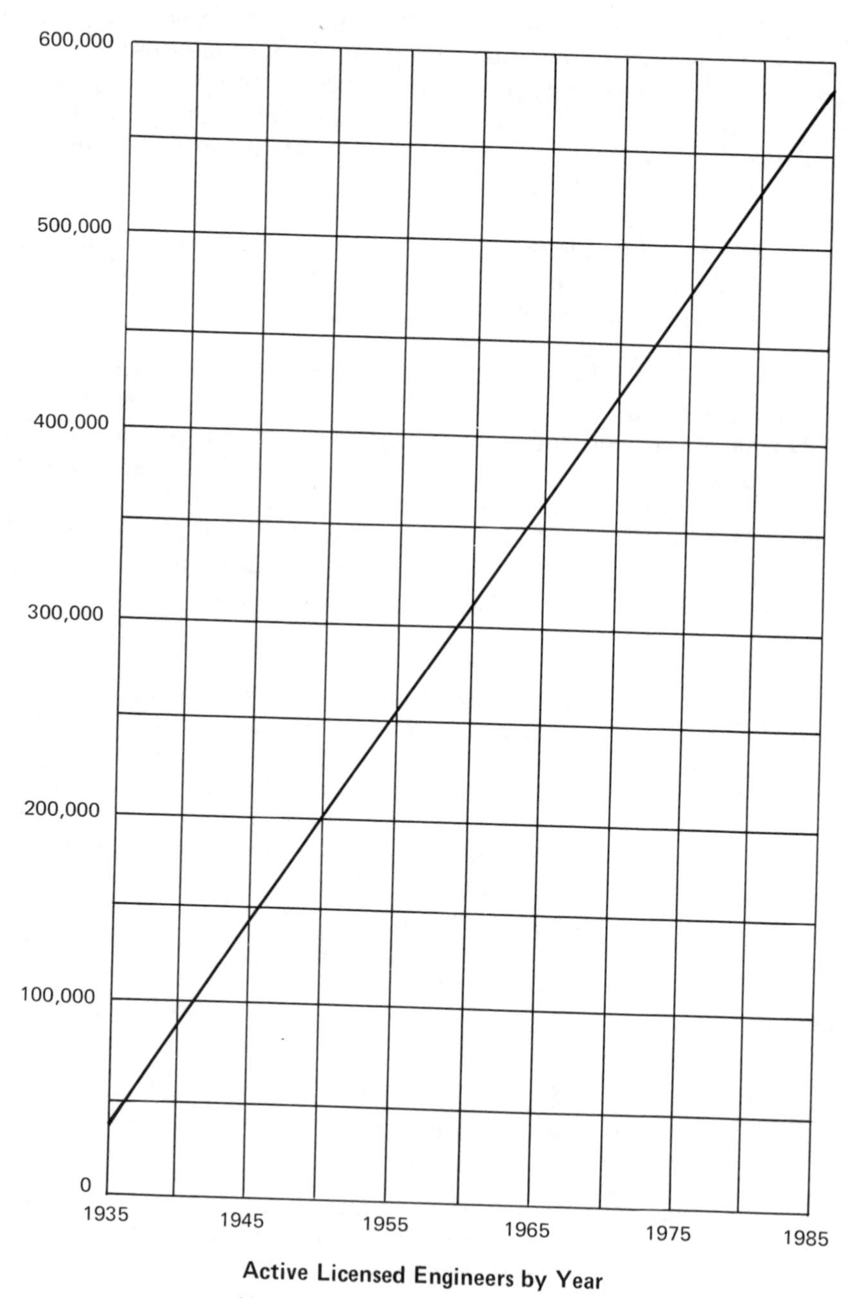

Active Licensed Engineers by Year
(All states and territories included.)

Figure 1

In addition to standardizing the registration examinations, NCEE has written a model law concerning registration of engineers. If adopted by all the states, this law would eliminate the many variations and exceptions resulting from the independent writing of each state's laws. In fact, the model law has helped to achieve some degree of uniformity, as many states have used it as a basis for their more recent statutes and procedures. However, adoption of uniform examinations has been more successful than acceptance of the model law.

The state boards administer examinations twice a year, typically in mid-April and late October. They also evaluate applications and their supporting documents, determine eligibility for taking the examinations, and (in some states) carry out interviews or oral examinations with applicants who pass the written examinations. Where NCEE examinations are used, NCEE does the scoring; where board-produced examinations are used, the board scores and grades the written examinations.

Names and addresses of all the state boards are listed in Appendix A.

LICENSING AND AREAS OF SPECIALIZATION

Registration, or licensing, may be defined as the legal recognition by a state or other jurisdiction of a person's qualification to practice engineering.

Most states follow the practice of single registration. This means that when you receive your license, you will acquire the title of "Professional Engineer," "Registered Professional Engineer," or the equivalent in your state. In this registration system, you are required to indicate your specialty when making your application. However, you are not limited to that specialty in answering examination questions, and you are not legally restricted to that area in your later practice. If your state follows this system, it will be up to you to determine whether you are qualified by training and experience to carry out any given project.

Other states register engineers by specialty area, for example, "Professional Civil Engineer," or "Professional Electrical Engineer." In those states, you must answer examination questions in your area, and you must restrict your professional activity to that area. In most states with specialty registration, it is possible for an engineer who meets the experience and examination requirements to register in more than one specialty area.

No matter which type of registration practice your state follows, for the record, you will be required to specify your area of specialty. Even if you are not required to stay within your area on the examination, the board may look for consistency between your education and experience and the area which you indicate as your specialty.

By far the largest number of candidates for licensure fall into one of the "Big Three" categories: civil, mechanical, and electrical engineering. However, there are sizeable numbers in other specialties as well, notably chemical engineering. In addition to these four areas, NCEE examinations have specialty options in manufacturing, ceramic, industrial, petroleum, agricultural, nuclear, sanitary, structural, and aeronautical/aerospace engineering. Some states offer their own examinations in areas not covered by the NCEE examinations. For instance, California offers examinations in metallurgical engineering and soils engineering. If your specialty is an unusual one, you should check with your state board on whether an examination is offered in that area, and if not, on how you should proceed.

In recent years, some professional engineering societies have begun to evolve a system of specialty certification parallel to the registration system. Certification is in no way a substitute for registration. Registration is legal recognition by the state of a person's engineering competence. Certification is recognition by a professional association of an individual's competence in a particular specialty within engineering. Many of the certification programs are in areas other than those covered by the standard examinations, and some of them are in narrow subspecialties. The main purpose of certification seems to be to define the specialties more precisely, and to provide a means of recognizing qualified specialists.

Many of the certification programs, in recognition of rapidly changing methods and developing technologies, require recertification every several years. Recertification typically depends on demonstration of professional development activities, such as further education or participation in technical meetings and conferences, or on retaking the certification test.

The recertification requirements of these professional societies have inspired several states to consider instituting a similar procedure for registered engineers. If this should occur in your state, maintaining your license would depend on showing evidence of your continuing professional development on a regular basis.

In addition to certification by individual professional societies, NCEE offers a program of national certification. This is not in any sense equivalent to national registration. All registration is done by the states. Rather, it is a program designed to assist the engineer who wishes to seek registration in more than one state. After review and verification of the required documentation, NCEE's Committee on National Engineering Certification may issue a certificate to the effect that the engineer meets standards more stringent than those of the individual states and jurisdictions. This certification may be of assistance to the engineer seeking registration outside the state of residence.

HOW TO BECOME LICENSED

What exactly is a licensed engineer? With the variety of different terms that are in use, some definitions may be helpful. Four terms may be considered synonymous: licensed engineer, professional engineer (PE), registered engineer (RE), and consulting engineer (CE). A person who has the right to use these titles is one who has successfully met all the requirements for education, experience, and examination performance set by the state granting the license or title.

Each state has its own requirements for registration, and these may change from time to time. Before applying for registration, you should get up-to-date information about the rules and requirements of your state. However, despite their differences in detail, all the states adhere to more or less the same fundamental principles in registration and licensing.

Each state requires some combination of education, experience, and test performance. In many states, experience may be substituted for education up to a certain point by counting a set number of years of experience for each year of education. The ratios and limits for this substitution vary from state to state.

Exemptions from Examination. There are certain limited conditions under which a person may be exempted from the usual examination. There are three ways in which exemptions may be granted.

The first such condition occurs at the beginning of registration in a new discipline, or when there has been a major change in the registration law. At such times, the new regulations may include a "grandfather" clause. Under a grandfather clause, engineers with a substantial amount of responsible experience in the field (usually nine years or more) may be given a period in which they are allowed to register without taking the examination. This is done in order to avoid unfairly putting these engineers' jobs into jeopardy. If those covered by a grandfather clause do not register within the allotted time, they will be required to follow the normal registration procedures.

A second type of exemption from the examination occurs only rarely. An engineer who has an engineering degree and who is judged to be eminent in a particular specialty area because of contributions to the field may be registered without examination in most states. This is simply a courtesy and a form of recognition extended by the state boards to the foremost experts and inventors in the various fields of engineering. It is not a practical way to circumvent the examinations—it is easier to pass the examinations than to document eminence in your field!

The third method by which a person may register without examination is through reciprocity or comity between the states. Reciprocity means that states which have substantially similar requirements for registration recognize each others' examinations as equivalent. Comity means that each state accords the same rights to citizens residing in other states that it accords its own citizens, within specified limits. Most states also have provisions under which the state board can evaluate the equivalence of licenses issued in foreign countries.

If you are licensed in one state, you can usually register without examination in other states. If you work in a company which does business in more than one state, registration by comity may be very useful to you.

The Examinations. In order to become a registered engineer, you must take and pass two eight-hour examinations, one in fundamentals of engineering and the other a more advanced examination covering specialized subject matter. Although in some states these two examinations may be taken back-to-back over two consecutive days, it is advisable to take them separately. The two examinations have different purposes and functions, and your progress towards registration will be more efficient if you follow the conventional sequence.

Some states require oral examinations of all candidates who have passed the written examinations. Other states interview or orally examine only those candidates whose educational background appears to be weak. You should check on your state's policy regarding oral examinations.

The first examination, called the Fundamentals of Engineering (FE) exam, is popularly known as the EIT (engineer-in-training) exam. Passing the EIT exam is the first step toward registration. It is the basis for taking the second, more advanced exam. The EIT exam may be taken while you are still enrolled in your engineering curriculum or shortly after graduation, before you have begun to accumulate engineering experience. It may also be taken at any later time. Passing this exam advances your professional status to that of engineer-in-training (EIT) or intern engineer (IE), which will help you to accumulate experience that is acceptable in qualifying for registration.

The second examination is known as the Principles and Practice or PE (professional engineering) exam. This exam can only be taken after the requirements for both education and experience have been met and after the EIT exam has been taken. In many states, the EIT exam must have been passed before the PE exam can be taken. Assuming that all other requirements have been met, success on this exam, and on the oral exam (if applicable), qualifies you for licensure.

THE EIT EXAM

Let us take a closer look at the EIT exam—who is eligible to take it, what the exam is like, and what to expect when you take it.

Who Can Take the EIT Exam? All the states set minimum qualifications for taking the EIT exam, although the details differ from state to state. The most usual basic requirement is graduation from a four-year program in engineering, basic sciences, or engineering technology which is accredited by the Accrediting Board for Engineering Technology (ABET). However, in most states, students in their last year or two in an accredited program are also eligible to take the EIT exam. Students in these programs are encouraged to take the exam as soon as they are eligible, and the schools themselves often assist students in preparing for the exam.

Graduates of non-accredited programs may be required to show a certain number of years of acceptable work experience before being allowed to take the exam. Applicants with little or no formal education in engineering must have graduated from high school and have several years of acceptable experience in order to establish eligibility. If your educational background is anything other than a four-year accredited program, check on your state's regulations regarding eligibility for the EIT exam.

A few states have age requirements for the EIT exam. The most common minimum age is 21. It is unusual for the age requirement to pose a problem, since virtually everyone who has graduated from a four-year program or who has accumulated the requisite experience in lieu of an accredited educational program will be older than the minimum age. Many states have dropped the age requirement, and others may be expected to do so in the future.

A few states, and most of the non-state jurisdictions, require U.S. citizenship or declaration of intention to become a citizen as a prerequisite to registration. However, most states do not have such a requirement, and those that do are being urged to eliminate it. If you are not a U.S. citizen, it would be advisable to learn whether your state is one of the few which still require citizenship.

Some states have residency requirements for the EIT exam. If you are attending an out-of-state school, it would be best to check on residency requirements in both your home state and the state in which you are attending school, and to find out whether you meet the criteria for state residence.

When is the Best Time to Take the EIT Exam? When should you take the EIT exam? The same answer holds for all candidates: as soon as possible! Questions on the EIT exam are chosen from the broad spectrum of basic engineering subjects, mathematics, and physical science. The earlier you take the exam, the more likely you are to be in command of this wide range of information.

If you are studying in, or have recently graduated from, an ABET-accredited engineering program, or have graduated from one recently, preparing for the EIT exam will be easier now than at any later time. The mathematics, science, and engineering that you have learned are still fresh in your mind, and you are still in the habit of studying and taking exams. It is greatly to your advantage not to delay much past your graduation.

If you are a student in, or a graduate of, a non-accredited program, you will have to meet your state's requirements for experience or further education before taking the exam. In this case, it is best to take a review course or follow a home-study program during the last several months of this required work, and to take the EIT exam as soon as you are eligible to do so. If you let more than the required amount of time elapse, you may run the risk of forgetting some of the material that you have studied in school.

If you are an engineer who has been working for a number of years in an exempt status, but believe that it would be to your advantage to register, you should apply for the next EIT examination for which you will qualify and begin to prepare for it immediately. You may be one of the many engineers who procrastinate with, "I'll do it as soon as I find time," or "I'll take it the next time around," or other similar lines. But the pressures of career and personal life make it progressively more difficult to find the time. Meanwhile, the broad knowledge of engineering you acquired in school fades or becomes outdated as the specific knowledge and skills required by your job occupy most of your attention. For you, as for more recent graduates, the sooner you begin to prepare, the easier it will be for you.

Dates and Locations of the Exam. The EIT exam is normally given twice a year, in mid-April and late October. In order to be eligible to take the exam, you must apply before the deadline set by the state board. In some states the deadline may be as much as four or five months before the exam is given. Although application for the EIT exam is much simpler than application for the PE exam, completing the forms and assembling the supporting materials may take some time. It is best to ask for application materials several weeks before the deadline and to file the application, including the application fee, as early as possible. Missing the application deadline means waiting until the next examination is given six months later.

When your application is accepted, you will be notified by the state board of the exact date, time, and location of the examination. Chances are that you will have to travel to another city, and you should arrange your time and budget accordingly. If there is any question about where you will take the exam, contact your state board for information.

Format and Content of the Exam. The EIT exam is an eight-hour exam, given in two four-hour sessions. In the morning session, there will be 140 questions drawn from thirteen subjects in engineering, mathematics, and physical sciences. You will be expected to answer all 140 questions. In the afternoon session, there will be 100 questions covering nine subjects. Of these, you will be asked to answer 70 questions covering six subjects. Although the number of questions differs, the two sessions carry equal weight.

In the morning session, you will be held responsible for all 140 questions. These questions are distributed among thirteen basic engineering-related areas, as shown in Table 1 below. Note that the number of questions allotted to each subject area may vary slightly from year to year, so that the numbers shown in the table are only approximate.

TABLE 1

	Subjects	Number of Questions
1.	Chemistry	10
2.	Computer Programming	8
3.	Dynamics	13
4.	Engineering Economics	6
5.	Electrical Circuits	18
6.	Fluid Mechanics	14
7.	Materials Science	6
8.	Mathematics	13
9.	Mathematical Modeling of Engineering Systems	8
10.	Mechanics of Materials	13
11.	Statics	13
12.	Structure of Matter	6
13.	Thermodynamics	14
TOTAL Number of Questions on Morning Section		140

The afternoon section contains 50 mandatory questions plus five sets of ten related questions. You must answer the 50 mandatory questions and choose two of the five sets for a total of 70 questions. Table 2 shows the distribution of questions on the afternoon section.

TABLE 2

	Required Subjects	Number of Questions
1.	Engineering Mechanics	15
2.	Mathematics	15
3.	Electrical Circuits	10
4.	Engineering Economics	10
	Additional Subjects (Choose Two)	**Number of Questions**
1.	Computer Programming	10
2.	Electronics and Electrical Machinery	10
3.	Fluid Mechanics	10
4.	Mechanics of Materials	10
5.	Thermodynamics/Heat Transfer	10
TOTAL Number of Questions on Afternoon Section (Examinee Answers 70)		100

The EIT exam is entirely composed of multiple-choice questions. For each item you will choose the most correct of the five alternative answers given, and mark the space containing the correct letter (A, B, C, D, or E) on the answer sheet provided. You are probably familiar with this type of exam if you have taken the Scholastic Aptitude Test, the Graduate Record Examination, or other similar large-scale exams.

As you can see from Tables 1 and 2, there are twice as many questions in the morning session as in the afternoon session, although the two sessions have equal weight. Each question from the afternoon section is worth twice as many points as each question from the morning section, because the afternoon questions are somewhat more difficult. The calculations required by the morning questions are likely to be quite simple, whereas some of the afternoon questions may require more complex solutions.

The best way to give you an idea of what the exam is like is to present a few sample questions. Here are five questions of the level of complexity that may be found on the morning section.

1. Which statement is true of magnetic flux lines?

   (A) They are continuous.
   (B) They diverge.
   (C) They may cross each other.
   (D) They terminate on electric charges.
   (E) They follow parabolic paths.

2. A square wave of alternating voltage with RMS amplitude of 20.0 volts is applied across a 4.00-ohm resistor. The RMS is

   (A) 1.77 A
   (B) 3.18 A
   (C) 5.00 A
   (D) 7.07 A
   (E) 10.0 A

3. Given the equation of a curve, differentiation can be used to determine all of the following EXCEPT the

   (A) slope
   (B) concavity

(C) location of inflection points
(D) number of inflection points
(E) area under the curve between points

4. The heat generated during combustion at constant pressure is equal to the

(A) change in chemical potential
(B) change in Gibb's function
(C) entropy change
(D) enthalpy of combustion
(E) enthalpy of formation

SELECT EITHER PROBLEM

5. (SI) A 40-gram bullet traveling with a horizontal speed of 900 meters per second collides with and is embedded in an 80-gram block of wood that is initially at rest on a horizontal frictionless surface. What is the speed of the block after impact?

(A) 900 m/s
(B) 600 m/s
(C) 450 m/s
(D) 300 m/s
(E) 150 m/s

6. (non-SI) A 4-ounce bullet traveling with a horizontal speed of 3,000 feet per second collides with and is embedded in an 8-ounce block of wood that is initially at rest on a horizontal frictionless surface. What is the speed of the block after impact?

(A) 3,000 ft/sec
(B) 2,000 ft/sec
(C) 1,500 ft/sec
(D) 1,000 ft/sec
(E) 500 ft/sec

The last sample question is given in two versions, one using metric or *systeme internationale* (SI) units and the other using English measurement units. This dual-version format is used in the EIT exam whenever calculations involve measurements of mass, length, or distance, and you are free to choose the measurement system you prefer to work with.

The questions on the afternoon section are organized as sets rather than as separate, independent questions. A diagram, sketch, or set of assumptions is presented, and a group of questions related to that information follows. In the first 50 items, which are mandatory, the number of related questions in a set ranges from three to six, with a total of ten questions for the category. The second 50 items consist of two sets of five questions in each of five specialty areas. You will choose two of these areas and answer all ten questions in each of those two areas. On your answer sheet you will indicate which two subjects are to be scored and which three subjects are not to be scored.

You may notice a form number or letter on your test booklet and answer sheet. (Two or more different forms are used to minimize the likelihood of cheating.) You will probably be asked to check that these numbers or letters are the same on both the booklet and the answer sheet. Since the order of questions is different on each form, it is important that you have the same form for both test booklet and answer sheet.

Statistics on Passing. You are probably wondering about your chances of passing the EIT exam. According to statistics published by NCEE, your chances are good. In recent examinations, about two-thirds of all those taking the exam have passed it.

The success rate varies according to whether or not the candidate has attended an ABET-accredited engineering program. In recent years, about 86% of all persons taking the exam have come from such programs. Students and graduates of accredited programs do a good deal better than the average—about 73% of them pass the exam. Students and graduates of non-accredited engineering programs do worse than the average, with about 57% passing.

Students and graduates of engineering technology programs do not perform as well, on the average, as those from engineering programs. About 35% of those from accredited engineering technology programs pass, and only about 27% of those from non-accredited engineering technology programs are successful.

There is a small group of examinees who qualify for the exam on the basis of experience rather than formal training. They include graduates of non-engineering programs, such as one of the physical sciences or computer science, as well as engineers who have had little or no formal training. About 35% of this group pass the EIT exam.

Among examinees from accredited engineering programs, the success rate varies according to major discipline. About four out of five candidates from these programs fall into the four major fields of civil, electrical, mechanical, and chemical engineering. Those from civil engineering perform at approximately the overall average rate, with about 66% passing. The other groups perform above the overall average, with 76% of those from electrical engineering and 81% of those from mechanical and chemical engineering passing the exam. (For details of passing rates, including other fields of engineering, you may wish to consult the *Candidate Performance Summary*, published annually by NCEE.)

Remember that these averages do not in any way determine your individual performance. It is your own talent and effort that will determine whether or not you will succeed. Even in the group with the lowest success rate (graduates of non-accredited engineering technology programs), one-fourth of the candidates pass the exam. If you are from that type of program, it is up to you and your efforts to determine whether you will be in the successful one-fourth or in the unsuccessful three-fourths. Likewise, if you are a member of the most successful group (from an accredited program in mechanical or chemical engineering), it is up to your efforts to determine whether you will be in the successful four-fifths or in the unsuccessful one-fifth. These statistics are presented only so that you can see how others with your educational background have fared on the exam.

How the Exam Is Scored. In scoring, only correct answers are counted. Incorrect answers and blanks are ignored, and there is no penalty for guessing.

In most states, the minimum passing grade is 70 on a scale of 100. However, this does not mean that you must get 70% of the items correct. Rather, the minimum passing score is standardized or scaled so that it is equal to a score of 70. For a grade of 70, you will have to answer about half of the questions correctly.

In terms of raw scores, the test has a possible 280 points—140 items from the morning section, and 70 double-weight items from the afternoon section. In recent years, the raw score which has been scaled to be equal to a grade of 70 has ranged from 135 to 142, and has usually been about 138, or just about half of the total possible points. The exam is monitored regularly so that the standardization can be done as precisely as possible. (For details on scaling or standardization of scores, see the

NCEE publication, *The Fundamentals of Engineering Examination: Procedures Used by NCEE to Set a Minimum Passing Score.*)

This method of scoring has important implications for you. First, and perhaps most important, you will pass the examination if you can answer about half of the questions correctly. Put another way, you can get about half the answers wrong and still pass. This means that you do not have to be an expert on every field of engineering. If you are well prepared in the subjects that you know best, you have an excellent chance of passing the exam.

A second important point is that, since guessing is not penalized, you can use a guessing strategy to improve your score. You can also be fairly sure of finishing the exam, at least in the sense of recording an answer for each question.

Another point is that there is no particular advantage in getting a high score. If you get a score of 85 or 90, you may be pleased or feel proud of yourself (and deservedly so). But in practical terms, a score of 70 and a score of 100 are equivalent: you pass with either score.

Exemption from the EIT Exam. In general, passing the EIT exam is the required first step toward registration as a professional engineer. However, under exceptional circumstances, some states will allow a candidate for registration to skip that first step and take the PE exam directly.

A candidate who would qualify for this kind of exception might be one who has an advanced degree, such as a Ph.D., in engineering, and who also has fifteen or more years of responsible experience as an engineer. Each state may have slightly different regulations regarding such exceptions, and some states may not permit skipping the EIT exam under any circumstances. If you believe that there is a chance that you might qualify for exemption from the EIT exam, check with your state board for more precise information.

What Happens If You Fail? If you fail the EIT exam, the old adage applies: if at first you don't succeed, try, try again. In most states you will be allowed to try again when the exam is given approximately six months later.

Your chances of passing the exam the second time around are better than the first time, because you will have a better idea of what to expect. Also, you can probably pinpoint your weak areas more precisely after the exam, and this will help you to direct your studying for the second try.

If you are not sure why you failed, you can ask your state board to let you see a copy of your examination. While you will not be able to keep a copy of the questions or your answers, you will be able to see what kinds of questions you tended to miss. You can then concentrate on making up your deficiencies before taking the exam again.

Do not forget that you do not have to know all areas equally well in order to pass. It may be a better strategy for you to improve your performance in areas that you know fairly well than to try to learn areas that you know almost nothing about.

You should contact your state board as soon as you learn of your failing score to find out how to reapply. Each state has slightly different regulations. You should learn whether you need to pay a new application fee, what forms need to be filled out, whether there is a time limit for reapplication, and whether there are any restrictions on when or how many times you can take the exam.

What Does It Mean If You Pass? When you pass the EIT exam, you may call yourself an engineer-in-training or an intern engineer, depending on your state. In this status, you can begin to accumulate the years of experience that you will need to qualify for the PE exam. Naturally, if you already have the required amount or more qualifying experience, you may apply to take the PE exam at the next opportunity. Passing the EIT exam does not make you a professional engineer, nor does it alone qualify you to take the PE exam. It is only the first step toward registration.

QUALIFYING EXPERIENCE

In the typical course of events, a candidate for registration will graduate from a four-year accredited program in engineering, take the EIT exam during the senior year, start work in an engineering position immediately after graduation, and begin to accumulate qualifying experience in order to take the PE exam at the earliest opportunity. The usual requirement is four years of qualifying experience.

It is not required that all your experience be accumulated after graduation. If you worked while going to school, and if the work fit the criteria for qualifying experience, you may qualify to take the exam less than four years after graduation. However, it may be difficult to prove that your pre-graduation work constituted true engineering experience.

If you are not a graduate of an accredited four-year engineering program, you will need more than four years of qualifying experience (usually eight to twelve, depending on your education) in order to be eligible for registration. Some states will not allow non-graduates to take the EIT exam at all, no matter how much experience they have. Your state board can provide information on the number of years of experience, if any, which may be substituted for each year of education.

While years of experience may be substituted for years of education, the reverse is not usually true. That is, the requirement of experience is not usually waived for candidates with advanced degrees, such as M.S. or Ph.D. degrees. However, different states have different regulations in this regard, and you should check with your state board for more specific information.

In order to be qualifying, experience must meet a number of criteria.

First, the experience should be in the major branch of engineering in which the candidate claims proficiency.

Second, the experience must be supervised. That is, it must take place under the ultimate responsibility of one or more qualified engineers.

Third, the experience must be of high quality, requiring the intern to develop technical skill and initiative in the application of engineering principles and sound judgment in reviewing such applications by others. The experience must be of such a nature that the intern develops the capacity to assume professional responsibility for engineering work.

Fourth, the experience must be broad enough in scope to provide the intern with a reasonably well rounded exposure to many facets of professional engineering. Along with highly specialized skill in a particular branch of engineering, the intern should acquire an acceptable level of competence in his or her basic engineering field, as well as in the accessory skills (e.g., cost estimation) necessary for adequate performance as a professional.

Finally, the experience must progress from relatively simple tasks with less responsibility to work of greater complexity involving higher levels of responsibility. As the level of complexity and responsibility increases, the intern should show evidence of increasing interest in broader engineering questions and continuing effort toward further professional development and advancement.

In assessing whether the candidate is sufficiently competent and responsible to be entrusted with or independently engage in engineering work, or to supervise engineering work, the board looks for evidence of independent decision making and assumption of personal accountability in design and application. In short, while the experience must be gained while under the supervision of qualified professionals, it must be professional in character.

Most of the functions which mark the engineer's work as professional revolve around various decisions which must be made in the course of a project. The comparison of and selection among alternatives for engineering work; the determination of design standards or methods; the selection or development of methods or materials to be used; the selection or development of testing techniques; the evaluation of test results; the evaluation of contractors' performances, methods, materials, and workmanship in relation to the integrity of the finished product; and the development and control of maintenance and operating procedures—all of these are examples of a professional level of functioning as an engineer.

NCEE has drawn up guidelines for the categorization of different types of experience as either professional or subprofessional. Limitations of space make it impossible to present the criteria in all branches of engineering here. However, the guidelines for mechanical engineering may give you a general sense of the distinction between professional and subprofessional work.

In mechanical engineering, the following types of experience may be considered professional: (1) the design of machines, machinery, heating, ventilating, and air conditioning equipment, power plants, power plant equipment, engines, tools and processes, mill or industrial layouts, and/or the supervision of the construction of any of these; (2) the development of industrial plants and processes, and/or consultation or contribution to such development; (3) operation, control, and testing of major mechanical installations, manufacturing plants, and power plants; (4) the writing of technical reports, manuals, and the like; (5) full-time teaching at an accredited college-level engineering school.

Mechanical engineering experience which is considered subprofessional would include the following: (1) construction and installation of machinery, heating, ventilating, and air conditioning equipment, and other mechanical structures; (2) operation of heating, ventilation, and air conditioning equipment, power plants, stationary machinery, mechanical manufacturing plants, and foundry and machine shops; (3) drafting, tracing, detailing, layout, and checking shop drawings; (4) designing tools, jigs, and fixtures; (5) recording data, routine computations under supervision, and inspection of materials; (6) maintenance and repair work; and (7) teaching as an assistant without full responsibility in an engineering program.

Some types of experience may be classified as either professional or subprofessional, according to the other types of work they are performed in conjunction with. If performed in conjunction with other professional work, they may qualify as professional experience. If they constitute the whole job, or are performed in conjunction with subprofessional work, they may not qualify. In mechanical engineering, these borderline tasks may include the following: (1) calculations of heat transfer, fluid transport, etc.; (2) the preparation of flow charts or logic diagrams; (3) the design of components and parts and the design of simple systems (e.g., fire protection, noise control, etc.); (4) reliability analyses; (5) installation of control, production, or environmental systems; (6) the laying out of plant equipment.

Sales work can be credited only if it can be conclusively demonstrated that engineering principles, knowledge, and skills were used in the work. Selection of equipment from a catalog or similar activities cannot be counted as engineering experience.

In general, the greater the complexity of the engineering work, and the greater the responsibility it entails, the more likely that it will be counted as professional experience. While you are working as an intern, it is important to seek opportunities to perform more complex work and to undertake greater responsibility, so that within a few years' time you will be operating at a fully professional level.

Documenting Your Experience. In applying to your state board for registration, you will have to document your experience and the fact that it meets the criteria outlined here. This documentation consists of two parts—your own statement of what you have done, and statements by your supervisor or supervisors, detailing the nature and extent of your experience. Most state boards will provide forms for both the candidate and the supervisor to use in documenting experience.

It is not unusual for experience to be disqualified because it has not been described in a way which could be evaluated by the board of examiners. Therefore, particularly with regard to describing intern experience, it is important that both you and your supervisor use the terminology and formulations that will be of greatest assistance to the state board.

Before filling in the portion of the application forms pertaining to experience, write a rough draft of what you want to say. Then review the draft for ambiguities and weak points. If possible, have someone who has experience and familiarity with the registration process review it also.

Let us look at some mistakes to avoid in documenting your experience.

Do not simply list job titles. No matter how impressive a title may sound, it must be accompanied by a detailed description of your duties and responsibilities in the position. This description must make clear the nature and extent of the engineering experience involved in the job.

Vague generalities and ambiguous phrases should be avoided. "I was involved in," "I worked on," "I was engaged in," and other similar phrases are uninformative unless they are followed by a specific description of duties. "I worked on the design of a cooling system for XYZ Factory," does not tell the examining board whether you worked as designer, draftsman, print coordinator, or something else entirely, or whether you did different jobs at different times during the project. Another type of vague phrase is "I was responsible for," or "I had full responsibility for." These phrases could also be used by an administrator who performed no engineering duties at all. It is much more useful to specify your duties precisely.

You should also avoid vague formulations regarding the amount of time you have spent performing each type of work. If you spent only a part of your time on a particular duty, indicate the percentage of your time that was devoted to that task. If you worked on a particular task on a full-time but intermittent basis, indicate the number of weeks or months that you spent on that activity.

You should not try to hide deficiencies in your experience through the use of vague, general language. It is better to wait until your experience is sufficient to qualify.

On the other hand, the application form is not the place for modesty. Do not assume that the full range of your duties, or the full extent of your responsibility, will be obvious from the job title or a brief summary. Failure to explain fully can lead to the rejection of your application. Go into detail, making sure that you give yourself credit for all that you have actually done. You may be surprised to find that a single job may encompass a number of engineering functions requiring many professional judgments. You should point out each of these functions and mention the types of judgments you were required to make, giving examples for major points.

In considering your application, the registration board must come to a decision as to whether your education and experience qualify you for registration. This means that the evaluation committee must be able to understand, evaluate, and verify the facts as you present them. A specific, detailed summary of your experience, written in clear, forceful language, will greatly increase your chances of qualifying for the PE exam.

If you have extensive experience, your employment history may provide more than enough years to qualify for the exam. If you are applying in a state which does not use the single registration system, you may prefer not to list all of your experience as qualifying for one particular license, especially if your experience has been multi-disciplinary. Years of experience held back in this manner may be used in a later application for a license in a different engineering field.

For example, suppose you have had ten years of experience in designing electromechanical control systems. If you can truthfully claim that your work has been 40% electrical and 60% mechanical, then listing only four years of experience in your application for an electrical engineering license will enable you to claim the other six years later in applying for a mechanical engineering license. If you claim all ten tears for the first license, you probably will not be able to use them later, since you will have, in effect, declared all of your experience to be in electrical engineering.

Other Requirements. Along with your experience record and verifications by your supervisor(s), you will need to have character references submitted to the board. Your packet of application materials will include information on how many letters of reference are required. Usually, character references must come from registered professional engineers who know you personally and can attest to your moral character and professional integrity. The information provided by your references may also be used to verify the type and duration of your work experience. Therefore, it is important to choose references who are familiar with your professional experience.

THE PE EXAM

Eligibility. Eligibility to take the PE exam is determined by the state registration board. All the qualifications that are necessary to take the EIT exam also apply to the PE exam. In addition, you must document four years (or more, depending on your education) of qualifying experience, and you must have the required number of character reference letters submitted. Your required years of intern experience must be complete as of the time of application, not the time of the exam. Only after the board has determined that you meet all of the requirements for registration, except for the examination, will you be eligible to take the exam.

When To Apply. Like the EIT exam, the PE exam is given in mid-April and late October in board-specified locations. Unlike the EIT exam, the application process is lengthy and complex. It is best to apply as soon as you believe you are eligible, but well in advance of the time that you actually expect to take the exam.

You should send for application materials four to six months in advance of the exam. State boards tend to move slowly, and it may take several weeks just to obtain your application. You will probably need at least one month, and more likely two, to prepare the application and assemble all the supporting materials. The board will then need time to consider your application and rule on your eligibility to take the exam. Consult your state board for application deadlines.

Format and Content of the PE Exam. The PE exam is a full-day, eight-hour exam which is divided into morning and afternoon sessions. Its purpose is to assess your capacity and readiness to assume responsible charge of engineering projects in your field of specialization. The PE exam consists entirely of problems or problem sets, the solutions to which are written into a blank exam booklet.

There are two basic options in the PE exam. One of them is called the *combined exam.* In this option, the exam booklet presents questions in the four major engineering fields (civil, including sanitary and structural;

chemical; electrical; and mechanical engineering), as well as in economic analysis. The other option, the specialties exam, consists of a different exam booklet with questions from nine other engineering disciplines, plus economic analysis. Table 3 summarizes the topics covered on each of the exam options.

TABLE 3

Topics on the PE Exam

Exam Option:	Combined Exam Option	Specialties Exam Option
TOPICS:	Chemical Civil Electrical Mechanical Economic Analysis	Aeronautical/ Aerospace Agricultural Ceramic Industrial Manufacturing Nuclear Petroleum Sanitary Structural Economic Analysis

The exam booklet option you take will be determined by the state board. If you claim specialization in one of the four major fields, you will take the combined exam. On the other hand, if you have a different specialty, you will most likely take the specialties exam. Also, some states administer examinations in specialties (e.g., metallurgical engineering) other than those recognized by NCEE. Whatever your specialty, you should learn the policy followed by your state board.

The states vary not only with regard to which version of the exam they require candidates to take, but also with regard to what types of questions they require or allow the candidates to answer within the exam. Some states require you to answer only questions within your specialty, but others may require or allow you to answer questions from other areas as well. Some states require that you answer questions on economic analysis, while others leave the choice up to you.

In view of this wide degree of variation in policies, it is to your advantage to learn your state board's policies as early as possible. Knowing what you will be required to do will allow you to plan your preparations much more efficiently.

Even if you will be allowed to stay within your field of specialization on the exam, inspection of sample exams will quickly reveal that they cover a broad range of material at a high level of complexity. The earlier you start your preparations, the better. You should allocate two or three nights a week for at least three months for review and study. Many engineers find that this amount of time is barely adequate, so you may want to plan either a longer or a more intensive study schedule. You should learn your state board's policies on examination areas a minimum of four months before the exam, so that you can gather the appropriate study materials and plan your study schedule ahead of time.

The PE exam is an open-book, free-response or essay type of examination in which partial credit may be given even though the final answer may be incorrect. Credit may be given for a correct approach to a solution, despite computational errors. There is a separate booklet of problems for morning and afternoon, and you will be provided with a blank workbook for each session.

In each session, you will be presented with ten problem sets in each of the specialty areas covered by your examination. You will choose and work four of them, following the instructions given with regard to specialty choices. You will be told whether you can, cannot, or must answer the economic analysis problem. The four problem sets that you work in each session are together worth a total of fifty points, so that the entire exam has a possible total of 100 points.

The style and level of difficulty of the PE exam can be demonstrated by presenting a few sample questions. The following are chosen from chemical, civil, electrical, and mechanical engineering.

Chemical Engineering

1. *SITUATION*: A new organic chemical is prepared by precipitation from an aqueous solution containing excess sulfuric acid. It is proposed to separate the solid from the solution by centrifugal filtration, wash the solid to remove most of the excess acid, and then dry it.

Pilot plant tests have shown that the centrifuge can be expected to give a residual mother liquor content of 0.08 pound per pound of dry product, and that a single stage of displacement washing, with 1 pound of water per pound of residual mother liquor, will give an 80% efficiency in removal of acid.

Assume that after each stage of washing, the residual mother liquor content remains 0.08 pound per pound of dry solid.

The design feed is:

2,500 lb/hr of solids
15,000 lb/hr of water
500 lb/hr of H_2SO_4

REQUIREMENTS:

(a) What product acidity, expressed as mass fraction H_2SO_4 in the dry product, can be expected from a single stage of washing with 1 pound of water per pound of residual mother liquor?

(b) If it is necessary to reduce the acidity to a 0.015 mass fraction H_2SO_4 (dry basis) or less, how many such washing stages are needed?

(c) At the end of the washings in (b), what is the acidity of the product (dry basis), and what is the H_2SO_4 content of the combined wash liquors?

2. *SITUATION*: A binary mixture of propane and n-butane exists at total pressure of 5 atmospheres absolute. Assume ideal behavior.

REQUIREMENT:

Calculate the end points and three intermediate points on the vapor-liquid equilibrium curve (y versus x).

Civil Engineering

3. *SITUATION:* Two cars separated by 120 feet are both traveling at 60 mph. The coefficient of friction between the roadway and skidding tires is .6. The trailing driver's reaction time is 1/2 second.

REQUIREMENTS:

(a) If the lead car hits a parked truck and comes to a stop, at what speed will the trailing car hit the lead car?

(b) How far will the trailing car skid after locking the brakes?

4. *SITUATION:* A W8x48 structural shape is placed on 5 foot centers to support a 4" thick slab as shown in the diagram. Each beam consists of a 25 foot simply supported section. The n-ratio is 8. Concrete weighs 150 pcf. Concrete strength is 4000 psi.

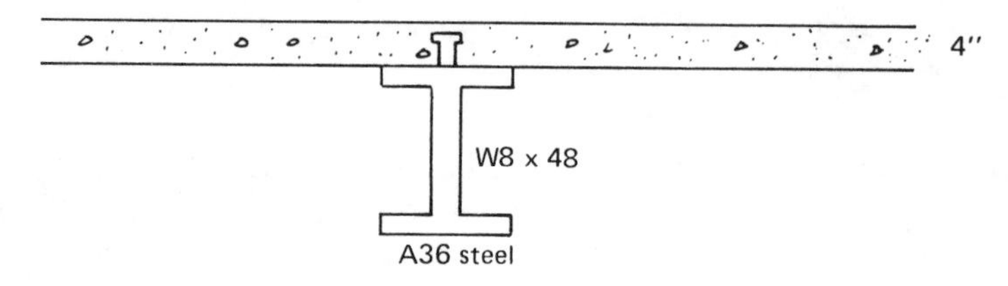

REQUIREMENTS:

(a) Find the maximum uniform live load that can be placed on the slab.

(b) What is the ultimate moment for this construction design?

Electrical Engineering

5. *SITUATION:* A shunt excited DC motor is rated 1150 rpm, 10 hp, 80 amps, and 40°F temperature rise. The motor drives a load for which the torque is to be constant regardless of speed.

REQUIREMENTS

Draw and identify the circuit elements for:

(a) a motor speed of 1300 rpm

(b) a motor speed of 800 rpm

6. *SITUATION:* An electrical supply of 20 volts ± 5% with 3 ohms internal resistance.

REQUIREMENTS:

(a) Design a Zener diode regulator to supply a variable 15 to 130 ohm load with a constant 15 volts DC.

(b) Sketch the complete circuit and specify all components.

Mechanical Engineering

7. *SITUATION:* An air conditioned building is to be kept at 76°F and 50% relative humidity when the ambient conditions are 96°F db and 76°F wb. The total heat load is 150,000 BTUH, with the sensible load being 80% of the total. 800 cfm of outside air are conditioned to 58°F before entering the building.

REQUIREMENTS:

(a) Find the required flow rate.

(b) Find the humidity in grains per pound of supply air.

(c) Find the tonnage rating of the air conditioner.

8. *SITUATION:* An orbiting spacecraft has an argon gas manipulator for deployment of a solar panel. The arm of the manipulator is connected to a 3-inch diameter piston with an 8-inch stroke requiring a 36 pound thrust in one direction only.

The argon gas is compressed to 1500 psig in a 0.1 ft^3 bottle. The bottle temperature is maintained at 380°R. There is a pressure regulator in the gas line. The ambient conditions at the time of launch are 14.7 psia and 540°R.

REQUIREMENT:

What will be the pressure in the gas bottle at the time of launch?

These problems may be considered representative of problems that may appear on the PE exam. They are practice-oriented, not textbook solutions or derivations. They require detailed written answers which explicitly indicate any assumptions made and clearly show all steps in the solution. While many of the questions on the exam are presented in a narrative format, many of them also include diagrams which contain necessary information.

Unlike the EIT exam, there are no alternate question forms using either SI units or English measurement units. Most questions use English units, but an occasional question will be presented in terms of SI units.

How the Exam Is Graded. There are important differences between the scoring methods used with the EIT exam and those used with the PE exam. Some of these differences are due to the different formats that are used on the two tests. For instance, partial credit is impossible with a multiple choice test like the EIT, while partial credit can be given on an essay test like the PE exam. Also, only one answer is correct for each item on the EIT exam, whereas problems on the PE exam (like problems in real-life engineering) may have more than one correct solution procedure.

In past years, grading on the PE exam was done according to a norm-referenced method, rather like the "curve" grading used in many university examinations. In this grading method, a certain percentage of examinees pass, regardless of the difficulty level of the exam or the absolute scores of the examinees.

Upon the recommendation of NCEE, a new system, called the criterion-referenced system, has been accepted by the states. With this new method, each problem worked is scored separately on a scale of 0 to 10 and is assigned a passing or failing mark. A score of 5 or less is considered a failing score on each individual problem.

To pass the exam, your performance must meet two criteria. First, you must have an average score of 6 or more over the eight problems that you work. That is, you must have a total of 48 points or more based on the raw score scale of 80. Second, you must have a passing score (6 or greater) on at least five of the problems that you work. Note that these two criteria are relatively independent of one another. It is possible to meet one of them and fail to meet the other. Failure to meet either of them means failure on the examination as a whole. This system has been adopted in order to ensure a minimum level of competence or qualification among those who pass the examination.

The score which is reported is not the raw score, but rather a standardized score on a scale of 100. The standardization is carried out in such a way that the same score reflects approximately the same level of performance from one exam to the other, despite a certain amount of fluctuation in the difficulty of the items or in the ability of the groups taking the exams.

Oral Examinations. In some states, a candidate for registration must appear before the state board for oral examination. This appearance is typically scheduled after the results of the written examination are known, and it is usually required only of those candidates who have passed the written exam. The oral examination is generally used to assess the candidate's judgment, professionalism, and poise under stress, rather than focusing entirely on technical competence. In this matter, as in many others, there is wide variation among the states. You should learn whether an oral exam is required in your state, and what types of questions are likely to be asked.

Statistics on Passing. Figures from a recent year indicate that roughly two-thirds of all candidates who take the exam pass it. It is much more difficult with the PE exam than with the EIT exam to make comparisons among candidates from different specialty areas. In some areas there are many candidates, and in others there are few or sometimes none. There may be little or no overlap in the questions answered by the different groups. The only statistical tendency that is worthy of mention is that graduates of ABET-accredited engineering programs do better, on the average, than candidates who come from other backgrounds. However, this tendency appears to be less pronounced than in the case of the EIT exam.

Exemption from the PE Exam. Certain categories of applicants may register as professional engineers without taking the standard set of examinations. In order of frequency, these are: (1) engineers who have already registered through the normal examination route in another state; (2) engineers who qualify under a "grandfather clause;" and (3) engineers who may be considered eminent, or who can claim "long established practice.'

Most states will grant temporary practice privileges to an engineer who is registered in another state, but these privileges tend to be very short term. If your business in another state will last longer than a few days, you should check with the host state's registration board for the regulations regarding registration by comity.

Many states have reciprocity or comity agreements with other states, recognizing one another's registration procedures as essentially equivalent. Registration by comity is a common practice. In recent years, approximately 40% of all engineering registrations in the United States have been non-resident (out of state) registrations.

One major exception to reciprocity practices is that states which require graduation from an accredited engineering program for registration may not recognize another state's license if it is based on test performance and experience in lieu of a degree. States which use the standard NCEE exams may or may not recognize licenses from those few states which administer their own examinations. Some other exceptions may result from differing state requirements regarding age, citizenship, length of experience, etc. However, the great majority of licensed applicants qualify for registration by comity in other states.

Thus, if you have a license in one state, based on the 16 hours of examination (EIT plus PE), and if you need a license to do business in another state, the second state will ordinarily waive the examination requirement. You should contact the host state's registration board for information on requirements, application procedures, necessary documentation, deadlines, and fees.

From time to time, as the practice of engineering becomes more differentiated and specialized, a state will recognize a new specialty in engineering, or redefine an old one, for examination and registration purposes. When this occurs, the new regulations usually include a grandfather clause. This clause gives practitioners with a specified amount of experience in the new discipline a grace period in which they may register in that specialty without examination. Those who register by means of grandfathering may be licensed in other fields, or they may be previously unregistered.

If your state is one which practices specialty registration rather than single registration, and if it does not currently recognize your field, you should follow any developments in the direction of recognition of your field. However, if there is no activity which would lead to recognition and a possible grandfather clause, it may be more to your advantage to register in a related field of engineering. Later, if your field should become recognized, you can make use of the grandfather clause to acquire a second license.

In a very few cases, part or all of the examination requirement for registration may be waived under a provision for licensing engineers who are "eminent" or who have "long established practice." Under this provision, engineers who have graduated from an accredited engineering program and who have many years of qualifying experience (usually 15 to 20 or more, depending on the state) may be considered for a waiver of the examination requirement. However, state boards do not automatically register all applicants who meet these degree and experience criteria; they look for a number of other superior qualifications as well.

Advanced degrees (M.S. or Ph.D), published technical papers, inventions, patents, copyrights, and contributions to engineering knowledge are among the signs of technical competence that a board will look for. Professional activity is also important, such as technical society membership, support of engineering education, development of training programs, work on scholarship committees, and continuing education and professional development. The recognition and esteem of other engineers is another important consideration. Teaching experience may be given credit. Evidence of civic responsibility and interest in community activities is also recognized.

A candidate who meets most or all of these criteria, and who has been in responsible charge of high-caliber engineering work for many years, may be registered without any examination at all. More likely, such a candidate may be given a "short test" to verify an acceptable level of engineering knowledge. The test may be either written or oral, or both.

Registration through eminence or long established practice is quite rare, and you should not consider it as an alternative unless you believe that you already possess the qualifications for this type of registration.

What If You Fail the PE Exam? Like the EIT exam, the PE exam can be repeated. If you are notified that you have failed the exam, you should contact your state board immediately for information on how to reapply. The board can tell you what forms must be used, what the reapplication fee is, and how long you must wait before you can apply again. In most states, you can take the next examination that is offered. However, if you have failed more than once, you may be required to wait a specific amount of time before trying again.

What Does It Mean If You Pass the PE Exam? If you pass the PE exam, including any oral exam that your state may require, you will be eligible to register as a professional engineer in your state. For registration, you will be required to pay a fee and file certain documents with the state board. The board will inform you about the specific procedures. As long as you pay your registration fees at the specified intervals and follow the regulations governing the professional practice of engineering, your registration will remain valid indefinitely.

PREPARING FOR THE EXAMS

Preparation for taking a major examination such as the EIT or PE exam calls for three main lines of effort: the gathering of information related to the exam, the organization of your time and activities, and studying the subject matter of the exam.

Gathering Information. It is hoped that this booklet will serve as a source of much of the information that you will need. However, no book can give you all the specific details that you must learn from your state board, such as application deadlines and fee amounts. Therefore, you should contact your state board as soon as you believe you are eligible to apply.

Do not forget that almost all state boards are part of a larger state bureaucracy and that they tend to be slow-acting. You may be able to get some information by telephone, but it may take several weeks for you to receive application materials by mail. This means that you should make your initial contact with the board far ahead of the application deadline.

Before beginning your application, make sure that you are aware of all the requirements, including all the forms and supporting materials. Read the instructions supplied by your state board carefully. Make a list, particularly if you are applying to take the PE exam, because the application process can be quite complex.

Getting Organized. Once you decide to take one of the examinations, you should establish a time frame and schedule for your preparation activities. Many experienced engineers recommend that you make your initial contact with the state board about six months before you expect to take the exam.

As noted in the previous section, you should allow a month or more just for receipt of application materials from the state board. It may not take that long, of course. Depending on the volume of work and other factors, the board may respond within a short time. If you do not receive any response within a month, you should call or visit the state board and inquire about the disposition of your request. It may have been misrouted or misfiled.

If you are applying to take the EIT exam, the application will be relatively simple. There will be forms to fill out, and you may have to provide documentation of your age, education, and possibly citizenship. You can probably take care of these requirements with only a few hours' work. Do not neglect the details, however. One state with a very simple application form recently reported that a large proportion of applications were rejected because applicants had neglected to follow such basic instructions as using a typewriter to fill out the form, or because they had failed to sign their names.

If you are applying to take the PE exam, the process is much more complicated. In addition to the documentation required for EIT applicants, you must document your experience and arrange for reference letters from professional engineers who are familiar with both your work experience and your character. Most applicants find that completing the application packet takes approximately a month of their time.

In preparing your application, remember that you are applying for recognition as a professional engineer. A professional-looking application packet, while it cannot substitute for qualifying experience and acceptable test performance, will surely make a good impression on the reviewing board.

Keep photocopies of all your application materials. Send the original application by certified mail and request a receipt of delivery. Keep your delivery receipt.

After you file your application, the board must rule on your eligibility to take the examination. This process may take a considerable amount of time. Do not wait for notification of the board's ruling to begin to plan your review and study time. If the ruling is slow in coming, you may find that there is not enough time left to study before the examination.

You should allocate at least three months at the rate of eight hours a week or more for review and study. This estimate is equally applicable to the EIT and the PE exams. Many engineers recommend even more study time. If you are working full time, as is likely, this is a significant investment of time.

Holding to such a schedule may entail a considerable reorganization of your usual evening and weekend hours, and it may require some adjustments on the part of your family and friends. If you are married, it may help to minimize family discontent if you discuss your study schedule fully with your spouse and children ahead of time. Let them know which days or evenings you will be unavailable so that they can plan family activities with your schedule in mind.

Establish a regular place to study, away from household traffic and noise. Keep all your books, notes, and other materials in that place for easy access while you study.

Set up a plan for the amount of time that you will devote to each topic. Set a target date for finishing your review of each topic, and stick to your schedule.

If you are studying for the EIT exam, it is best to spend more time on topics that you know moderately well than on topics that you know very well or not at all. If you are studying for the PE exam, give yourself enough time on each topic to master it well. Do not neglect to allot time for subareas within your major field that you are less familiar with. It is possible that all the exam questions will fall outside your narrow subspecialty.

Plan to finish your studies several days ahead of the exam, leaving a few days for overall review. Set aside at least one day before you leave for the exam location for relaxation or for some leisure activity which you enjoy. Last-minute cramming is more likely to raise your level of anxiety than your level of knowledge.

If you live far from the examination site, allow adequate time for travel. Make arrangements or reservations for transportation and accommodations well in advance of the exam date. Even if you live in the same city or metropolitan area, but are some distance from the exam site, you may wish to consider staying in a more conveniently located hotel the night before the exam.

Studying for the Exams. In planning your strategy for exam preparation, first take stock of your situation and the resources that are available to you.

Which exam are you planning to take? The EIT exam covers a broad spectrum of engineering knowledge, which means that you must review basic science, mathematics, and many different fields in engineering. The PE exam requires more detailed knowledge of fewer subjects, which means that your study should be more concentrated in the areas you know best.

What is your educational background? If you are currently a student in an accredited engineering program, chances are that your school provides a review course or other assistance in preparing for the EIT exam. If you are already out of school, you will have to arrange your own study program.

Are review courses available in your area? Some colleges and universities offer extension courses designed for EIT candidates and for PE candidates in some fields. State chapters of NSPE and some technical societies sometimes give refresher courses. In some areas, organizations such as the Professional Engineering Institute and engineering clubs offer review courses. For information on courses available in your state, contact NSPE in Washington, D.C., as well as your state professional engineering society. Addresses of state professional engineering societies are listed in Appendix C.

Particularly if you have been out of school for some time, or if you are not a graduate of an accredited engineering program, a review or refresher course can be extremely helpful. It not only provides actual instruction in the subject matter, but it also helps to structure your time and gives you regular practice in the types of problems that are likely to appear on the examination. If you are a recent graduate, a review course is less likely to be necessary.

Before enrolling in a course, you should check on several points to make sure that it will be worth your while. Does the course cover the subject areas that you need to review? Are the instructors registered engineers or other reputable experts? Are there any prerequisites for the course? (A good course will probably require you at least to be eligible to take the EIT exam.) Does any engineering organization or state education agency recognize, endorse, or accredit the course? How does the cost compare with other such courses?

If there is no review course offered in your area, you may wish to consider a correspondence or home-study course. Although there is no classroom instruction, a correspondence course could help in directing your attention to all the necessary subjects and in keeping you on a study schedule. National and state offices of NSPE can supply you with lists of some of the home-study courses that are available. As with classroom courses, you should check on the reputability and usefulness of any home study course before enrolling.

If you prefer not to enroll in a course of any kind, there are many books which have been written specifically to assist in preparing for the exams. A list of review manuals published by Professional Publications, Inc. may be found following the title page of this book. Upon request, NSPE will provide a bibliography of publications that may be helpful in preparation for the exams.

For each basic topic that you study, prepare a quick reference or review sheet. On one sheet of paper, summarize the fundamental concepts, relationships, equations, and formulae related to the topic. Collect all these information sheets into a binder for quick reference during the exam.

Along with review manuals, practice exams can be extremely useful. A major portion of your study time should be devoted to working sample problems. Merely reading the problems is not sufficient. A large part of what the exam tests is your ability to grasp a problem quickly and move rapidly to a solution. The speed you will need in the exam can only be attained through practice.

Most review manuals contain practice questions. There are also books which consist entirely of practice exams. NCEE and NSPE will both provide sample exams upon request. Since working practice problems is one of the most important components of your exam preparation, you should make use of as many sources of practice questions as possible.

You will gain the greatest benefit from the sample exams if you take them under fairly realistic conditions. Do not preview the questions, but wait until you have finished your initial review of all topics, working many practice problems in each topic. Then take a sample exam, observing the indicated time limit and following the rules that will apply in the actual exam. When you score the exam, make note of your areas of weakness, and go back for further review in those areas. Following this remedial work, take a second sample exam, again under realistic conditions. You will be able to gauge your progress quite accurately if you proceed in this manner.

Final Preparations. If you have followed all the recommendations of this section, you will approach the examination with a high probability of success. All that remains is to plan for your day (or two days) in the examination room and to adopt an attitude of confidence based on your thorough preparation. The next sections will help you in this final planning phase.

EXAM-TAKING TIPS AND STRATEGY

Last-Minute Considerations. Plan your time so that you will arrive at the exam room at least half an hour before the examination starts, and avoid the extra tension of worrying about being late. This will improve your chances of finding a convenient parking space, and will give you adequate time for any site, room, or seating changes that might be necessary. Do not forget that you could be refused entrance to the exam if you are not there on time.

If the exam is being held far from your home, get a room as near the location of the exam as possible. If you are not familiar with the location, go there ahead of time. Learn how long it will take to get there in early-morning traffic from your hotel, where you can park, how long it will take to get a taxi, how to get to the actual exam room, where the restrooms are, and any other details that could create problems for you on the day of the exam.

Take local street maps with you, and be prepared for the local weather. Take a raincoat, hat, boots, gloves, and umbrella.

Whether you are spending the night before the exam at home, in a hotel room, or with friends, don't try to study. Stay away from late-night entertainment as well. Your performance will benefit more from a good night's sleep than from anything else.

On the day of the exam, if you must park in a time-limited zone, leave a note to the parking patrol on your windshield explaining your situation. However, it is better to avoid this situation if possible. If there are no adequate parking facilities near the exam site, you may prefer to take a taxi.

In making arrangements for babysitters, return transportation, etc., allow for a delayed completion.

What to Take to the Exam. Your examination kit will be rather large and extensive, and you should prepare it the day before the exam. Here are four checklists of things you should take to the examination.

First, you will want to guarantee your admission to the examination. Take with you the following:

- ☐ photographic identification (e.g., your driver's license)
- ☐ letter from the state board admitting you to the examination
- ☐ your copy of your application and delivery receipt

Since the exam is an open-book exam, you will want to take useful references. You will not be allowed any loose paper; all notes and papers should be gathered into binders. Make sure you have the following with you:

- ☐ your review manual(s)
- ☐ your course notes, if any, in a three-ring binder
- ☐ book(s) of relevant mathematical tables
- ☐ a dictionary of scientific terms
- ☐ a good standard English dictionary
- ☐ a cardboard box cut to fit your references

You will be provided with answer sheets or an answer booklet, but you will have to provide all other materials. The following are recommended:

- ☐ several sharpened #2 pencils (necessary for machine-scored answer sheets; the machine cannot read anything else)
- ☐ mechanical (twist-to-advance) pencils
- ☐ extra pencil leads
- ☐ a pencil sharpener
- ☐ a large soft eraser that will erase completely without tearing the paper
- ☐ colored pencils or a highlighter pen (for PE exam)
- ☐ graph paper in a binder
- ☐ scratch paper in a binder
- ☐ straight-edge, ruler, compass, and protractor
- ☐ calculator (any type, but it must operate silently), and a spare
- ☐ spare calculator batteries or battery pack
- ☐ battery charger and 20' extension cord
- ☐ desk expander—if you are taking the exam in auditorium chairs with tiny fold-up writing surfaces, you should bring a long, wide board to place across the arm rests
- ☐ watch
- ☐ scissors, stapler, and staple puller
- ☐ transparent tape and masking tape

Finally, remembering that the exam is a long and grueling one, you should arrange to be as comfortable as possible. You will find the following helpful:

- ☐ a light lunch for the noon break
- ☐ thermos filled with milk or hot chocolate
- ☐ snacks such as cheese, raisins, nuts, or trail mix
- ☐ a chair cushion—a large, thick bath mat works well
- ☐ earplugs
- ☐ extra prescription glasses, if you wear them
- ☐ sunglasses
- ☐ aspirin or other headache remedy
- ☐ prescription medicines, if you are taking them
- ☐ travel pack of Kleenex
- ☐ a light, comfortable sweater
- ☐ comfortable shoes or slippers for the exam room
- ☐ $2.00 or more in change
- ☐ battery-powered desk lamp
- ☐ construction paper for stopping drafts and sunlight
- ☐ a wire coat hanger

What NOT to Take to the Exam. Certain items are forbidden in the exam room. Make sure that you do not include them in your examination kit.

In some states, you may not take into the exam sample questions and/or their solutions. (This includes Schaum's Outlines, NCEE's publications, and review books which are purely problem oriented.) You may not use a calculator which makes noise in its operation. You may not have any loose papers or notes. However, there are no rules against notes in the margins of books, or against reference materials and other papers in binders.

Do not burden yourself with unnecessary books and materials. You may see some candidates arrive with what appears to be their entire engineering libraries. This is not only unnecessary, it is also inefficient. There is no time during the exam to use references extensively, and you may not be given enough space to keep so many books with you. Take only the few references that you know best and use the most. They should be well marked or highlighted, and index tabs should be added for quick reference. Other references will either remain unused or will constitute a time-consuming distraction.

TAKING THE EXAMINATION

After the many months of preparation, you finally find yourself in the examination room, about to begin the exam. What can you expect during the exam? How can you make best use of your time? What strategies can you use to maximize your score? What happens if you get sick? Answers to these questions and more can be found in this section.

The Examination Setting and Schedule. Chances are that you will take the examination in a large school auditorium or other large lecture hall. It is also possible that you will take it in a smaller classroom or similar setting. Be prepared for virtually any conditions of heating and lighting—notice that the checklist includes both sunglasses and a battery-powered lamp, for instance. Do not forget that morning and afternoon lighting conditions may be quite different, and that you will be there through the whole day.

Before entering, you will have to identify yourself, presenting both the state board's letter admitting you to the exam and another photograph-bearing piece of identification. If seating is assigned, you will be told where to sit.

Be prepared for delays in starting the exam. A magazine or other light reading matter may help to pass the time while you are waiting.

When the exam booklets and answer sheets or booklets are distributed, one of the proctors will give instructions. Do not distract yourself by leafing through the exam booklet while instructions are being given, but listen carefully. You will be told how to use the answer sheet or booklet. You will also be told which questions you must answer, which questions are optional and how many of them you must answer, and which questions you may not answer. If you do not follow the instructions precisely, you run the risk of having your examination disqualified.

The proctors, who usually are not engineers, are there to ensure that the exam is administered correctly. They may answer questions regarding test administration or personal emergencies, but they cannot answer substantive questions about the exam. They cannot define terms for you, or tell you what a particular question means. It will only waste your time to try to get such information from the proctors.

The morning session will last for four hours. At the end of the session, all exam booklets and answer sheets or answer booklets will be collected.

There will be a break for lunch, followed by the afternoon session. At the beginning of the afternoon session, new exam booklets and answer sheets or booklets will be distributed. You will be given instructions for the session. In the EIT exam, the format of the afternoon session is different from that of the morning session, and you should pay close attention to the instructions. You will be given four hours to work on the afternoon section of the exam.

Strategies for the EIT Exam. A strategy for maximizing your score on the EIT exam includes two main elements: you must manage your time well, and you must guess effectively on items that you do not know.

The morning section comprises 140 questions which must be answered in four hours, or 240 minutes. This means that you have 1.7 minutes (one minute and 42 seconds) per question, on the average. Do not waste time by skimming through the questions, trying to find the easier items. Start at the beginning and work your way through systematically.

In terms of difficulty, the questions are likely to fall into four categories: (1) questions that you can answer immediately, or with minimal calculations; (2) questions that you feel sure you can answer, but that will require more time for calculations (although none of the questions require elaborate calculations); (3) questions that you are not sure of, but believe you can make an educated guess at; (4) questions which are entirely outside your range of competence. You will make most effective use of your time if you treat these different categories of questions differently.

On your first pass through the exam, answer all the questions in the first category and label the rest with some symbol (such as 2, 3, 4) that will remind you which categories they fell into. Then go back and do the questions in Category 4.

This suggestion may sound unreasonable to you, because your first instinct would be to return to the items that you know something about. However, you will leave yourself more time at the end for calculations if you get the guesswork out of the way first.

Some testing experts recommend choosing one letter alternative (A, B, C, D, or E) and marking that alternative for each of the items for which you are making a random guess. Assuming that correct answers are randomly distributed among the five alternatives, choosing the same alternative each time you have to guess will give you approximately 20%

correct over all the guessed items. Randomly marking a different alternative each time may reduce the percentage of correct guesses.

Having finished with the pure guess items, which you should have completed in a short time, return to the items on which you believe you can make an educated guess (Category 3). Move as quickly as you can through these items. For some of them, you will be able to isolate the correct answer by eliminating the other answers on grounds of irrelevance or illogic, while for others you will be left with two or three alternatives among which you will have to guess. If you really have no basis for a choice in these cases, quickly make a random guess and move on. At least you will have improved your chance of a correct answer from 1 in 5 to 1 in 3, or even 1 in 2.

At this point, check your time and the number of questions you have left to answer. If you have followed the suggestions given here, you should have used less than 1.7 minutes per item for the previous categories, leaving you somewhat more than that for each of the items in Category 2.

Do not spend too much time on any one question. In the morning session, each item counts for only one point. If you cannot work out a problem in a reasonable amount of time, make a guess and move on.

Following this strategy, you will probably finish the exam on time. If, however, you still have a number of items left in the last few minutes, finish them all by random guessing. You will probably gain more points by guessing than by working one or two more problems. Remember that there is no penalty for guessing. Make sure that you have recorded an answer for every question.

The afternoon session requires a similar strategy, but with some additional considerations. There are fewer questions in the afternoon session, each of them worth two points, and they are divided between mandatory and optional sections. With 70 questions in four hours, you will theoretically have an average of 3.4 minutes to spend on each item.

In reality, you will have less time than that, because you will need to spend time reading the optional questions and deciding which two of the five optional subjects you will choose. You may go into the exam believing that you will choose the two subjects which are closest to your area of expertise. However, after you see the test questions, you may decide to change at least one of your choices.

It may help you in keeping a steady pace if you mentally divide the afternoon session into two segments, one for the 50 mandatory questions and the other for the 20 questions in additional subjects. If you prefer to approach the afternoon session in this way, then you should set yourself a deadline for finishing the required 50 questions. This should be a little less than three hours from the beginning of the session. This will leave you approximately one hour for questions in the two additional subjects.

In answering the questions, follow the same strategy as for the morning session. Answer first the questions that you can solve quickly, then those for which you must make random guesses, then those for which you can make educated guesses, and finally those which require longer calculations. Follow this sequence separately for the mandatory and the additional subjects, if you are dealing with them in different time segments.

The answer sheet for the afternoon session will include boxes marked "score" and "do not score" for each of the optional subjects. Be sure to mark these boxes correctly for each subject.

If you finish either session of the exam with time to spare, as some people do, you may be allowed to leave the examination hall early. However, it is more to your advantage to use the time to check over your exam. Remember that one item can make the difference between passing and failing.

Take some of the time left to go back and check your calculations, especially those which may have required unit conversions. For items which could be worked in either SI or English units, you can double-check your answers by reworking the problem in the other unit system. The answer choice (A, B, C, D, or E) will be the same for both versions of the question.

Use your extra time to resolve doubts about particular items by looking up relevant material in your reference books and notes.

If there is still time, scan the items on which you guessed. While working on other items, you may have remembered information which would allow you to improve your guesses or even find the correct answer for some of these questions.

Since you will be skipping items on each pass through the exam, be especially careful to record your answers in the correct blanks on the answer sheet. If you have time left at the end, do a spot check to see that you have recorded your answers in the intended places, and that you have answered every item.

If you need to erase an answer, be sure to erase completely. The optical scanner which scores the exams will pick up incomplete erasures and stray pencil marks. If there are two answers (including unintended pencil marks) for any item, that item will be scored as incorrect. Before turning in your answer sheet, check it over for incomplete erasures, stray pencil marks, and blanks. There should be one and only one response to each question.

Strategies for the PE Exam. Maximizing your score on the PE exam requires good organization and time management. Guessing will be of no help, since the exam is entirely in an essay format.

Good organization has two aspects in the P.E. exam: you must choose your problems well, and you must present your solutions in an effective manner. Let us look at these two aspects separately.

At the beginning of the exam period, you will be given instructions regarding your selection of problems. These instructions should reiterate what you have already learned from your state board regarding your choice of questions. Depending on your state regulations, you may have more or less latitude in choosing which problems to solve.

If you are allowed to choose questions outside your major specialty, quickly review all the questions. You may find one that appears to belong to another specialty, but actually is closer to your own area than the questions that are ostensibly in your specialty. Within your limits of choice, pick the problems that you are most confident that you can solve.

You will solve eight problems in eight hours (four in the morning and four in the afternoon), so that you will have approximately one hour to spend on each problem, minus the time you spend reading and choosing problems. It is best to budget your time so that you do not spend more than an hour on any given problem. Work the easier problems first, to give yourself more time for more difficult questions later.

Once you have chosen a problem, you must not only solve it, but present its solution in a way that will gain you maximum credit. The first step towards successful solution is to read the whole problem carefully before you begin working on it. Be sure that you know both the assumptions that you must make and the question(s) that you must answer. You will lose valuable time if you get halfway into a solution, only to discover that you have ignored one of the assumptions, or that you are answering a different question than the one that is actually being asked.

Many problems will include extraneous information, and some may not have all the information that you need, so that you must make additional assumptions. You will proceed more efficiently toward a solution if you strip the problem down to its bare essentials. Do not waste time copying the problem or redrawing figures into your answer booklet. On the other hand, it may help to jot down the main points.

Keep your written presentation as well-organized and neat as you can. You will be asked to write on only the right-hand pages of the answer booklet. Use the left-hand pages as scratch paper, or use paper in one of your bound notebooks. Write neatly, and label all diagrams and drawings legibly. Neatness is no substitute for a correct solution, but the examining committee must be able to read your solution in order to give you credit.

Begin with a diagram or sketch, if appropriate, and list the physical principles which are related to the problem. If possible, write them as formulae or equations. List the variables given in the problem statement, along with their values. Determine whether any data are missing from the problem statement; you may need to consult tables, or make additional assumptions. If you do make assumptions, state them clearly and label them as assumptions.

If necessary, carry out algebraic solutions before substituting numerical values into the equations. When you do substitute the values given, make sure that they are in compatible units. Be careful with unit conversions (e.g., feet into inches, grams into kilograms, etc.).

When you calculate the answer, check it against common sense: Is the value reasonable, given the question? Label the answer with the appropriate unit. An unlabeled number does not constitute a complete answer. For instance, if the question asks for speed, give the answer in miles per hour, feet per second, or other appropriate unit.

A correct answer is not sufficient by itself. You must show each step in your solution, and specify the assumptions you are using. Do not leave out steps because they seem obvious to you. The person grading the exam will have no way to distinguish between the things you left out because you understood them too well and the things you left out because you failed to understand them. Similarly, solutions obtained largely by the use of preprogrammed calculator modules may be worthless on the exam.

Even if your calculations are wrong, you can get partial credit if you show that you are using the correct method to solve the problem. It is desirable, of course, to find the correct answer. If you have time at the end, check your calculations to make sure that you have used the correct numbers, signs, and units throughout.

Many of the problems, in addition to diagrams and calculations, require some verbal exposition. Write so that your meaning is clear and unambiguous. Check back over what you have written with a reader's eye: Have you said what you meant to say? If necessary, do a little editing to clarify your answers.

Because of time limitations, you will probably not be able to present your solution as elegantly as you would wish. To help the reader follow your line of thought, set off major points by underlining with a colored pencil or with a highlighter pen.

Managing Anxiety During the Exam. Feelings of stress and anxiety are very common during the exam, and may even be considered desirable, up to a certain point. You are likely to do better if you are somewhat keyed up than if you are completely relaxed. On the other hand, a high level of anxiety, or feelings of panic, will make it more difficult for you to keep your mind on your work. You will probably have no trouble getting keyed up enough to do well—the external stress of the exam virtually guarantees that you will not be overly complacent. It is more likely that you will have trouble keeping your anxiety down to a manageable level.

The most important factor in managing test anxiety is being well prepared ahead of time. This includes not only thorough study of the subject matter and exam practice under realistic conditions, but also knowing what to expect during the exam. Simply reading this book should eliminate many possible sources of anxiety.

No matter how well prepared you are, however, you may find yourself getting too anxious, feeling panicked, or not being able to concentrate on the questions. This may happen if, for example, you see that you are falling behind your schedule, or if you encounter a long series of questions whose answers you can only guess at. You should deal with this kind of situation with some simple mental exercises.

If you regularly practice some type of relaxation technique, take a minute or two to use the technique and calm yourself down. If you do not normally practice any such technique, take a few slow, deep breaths and remind yourself of the following facts.

1. You do not need to know everything to pass the exam.

2. Each question has only a limited number of points.

3. Thinking about yourself (how you feel, what will happen if you fail, etc.) prevents you from thinking about how to solve the problems.

4. It is an open book exam. If you cannot remember where to find a particular piece of information, try your dictionary. The definition of a term may make it clear how to solve the problem.

5. You will feel better if you move on to something easier. Once you feel calmer, it will be easier to go back and solve the problem that precipitated the panic.

If you have a long history of extreme test anxiety, you may want to consider seeing a counselor or therapist. Most people with test anxiety can be helped greatly in a short period of time. If you simply have the more common tendency to panic when you encounter difficulties on a test, being mentally prepared and practicing the recommendations given here should enable you to manage your anxiety successfully.

Unexpected Problems. Despite the most careful preparations, it is possible that some unforeseen problem will crop up during the examination. This section will give some suggestions for dealing with the unexpected.

For instance, what should you do if you become ill or suffer some other personal emergency during the examination? If it should become impossible for you to finish the examination because of illness, write "ILL—INCOMPLETE" across your answer sheet or booklet. When you turn in your test booklet and answer sheet, call the proctor's attention to the fact that you are leaving the exam incomplete because of illness.

Taking this precaution will prevent your exam being graded and possibly being given a failing score. In states which impose a waiting period or a remedial education requirement on candidates whose scores fall below a certain point before allowing them to take the examination again, the distinction between an incomplete exam and a failed one is especially important.

It is quite possible that there may be delays in starting one or the other session of the exam. Do not be concerned. In such a case, adjustments will be made so that you will have the full four hours to work.

In the EIT exam, you may encounter a question for which you believe that none of the alternative answers is correct, or for which you consider two of the alternatives to be correct. Because the exam is prepared by a group of experienced experts, and checked and rechecked again before publication, it is highly unlikely that the question is actually invalid. Nevertheless, it is not impossible for a typographical error to change the meaning of an otherwise valid problem.

Such a question will almost certainly be discovered during the scoring process. If there is no correct answer among the alternatives, the question will be eliminated from scoring. If there are two correct answers, either of them will be counted as correct, or the item will be eliminated. In either case, your score will not be affected adversely.

Your best course of action is to answer the item to the best of your ability, without wasting too much time on it. Make a random guess if necessary. Mark only one alternative, even if you believe that two alternatives are correct.

If you discover, at any point during the exam, that there is any defect (such as missing pages) in your exam booklet, notify a proctor immediately.

Cheating. Many precautions are taken against the possibility of cheating during the exam. Candidates are seated so that it is difficult for them to see one another's papers, and alternate forms of the exam are given to those who are seated nearest one another.

You may become aware that another candidate is trying to see your paper. Since the examination is given in alternate forms, it is likely that your answers will be worse than useless to anyone else. Nevertheless, you should keep your answer sheet directly in front of you, so that it is difficult for anyone else to see your answers, and so that you will not be suspected of aiding someone else's attempts at cheating.

If a proctor suspects an examinee of cheating, that person may be expelled from the exam immediately. It goes without saying that you must not cheat in any way. For your own protection, you should also avoid any behavior that would give even the appearance that you are trying to see someone else's paper, or attempting to exchange information with another person. A proctor's attention will immediately be drawn by talking, by the exchange of any items, such as paper, erasers, or personal supplies, or by any unusual posture or behavior. Make sure that you bring with you everything that you will need, and keep your eyes on your exam paper.

HINTS FOR THE ORAL EXAM

Not every state requires an oral examination. Some states require every candidate for a PE license to appear for an oral exam, while others require it only under special circumstances. You should learn your state's policy regarding oral exams.

The nature of the oral exam may vary according to its purpose. Where the exam is being used to screen candidates with borderline performance on the written exam, it is likely that questions will tend to be content oriented. On the other hand, where an oral exam is given to all candidates who have passed the written PE exam, the questions are likely to be more practice oriented. In either case, the examining committee will be trying to assess the professional qualification, judgment, and poise of the candidate.

If you must appear for an oral exam, your ability to answer questions regarding engineering knowledge and judgment will be the most important factor in your success. However, your self-presentation may also play a role in the committee's decision. If you dress neatly, speak clearly, and assume a matter-of-fact, professional manner, you will make a favorable impression on the examining committee.

COMPLETING YOUR REGISTRATION

Along with notification that you have passed your exams, your state board will inform you of the remaining steps you must take in order to become fully registered as a professional engineer. A registration fee must be paid. (Renewal of registration fees are also charged on a regular basis.)

After paying the registration fee, you will be issued a registration card and a registration certificate. You will also be authorized to buy a state-approved seal and stamp, to be used in certifying your professional work. You will probably receive a list of all professional engineers in your state.

WORKING AS A PROFESSIONAL ENGINEER

If you did not receive it at the time of application, you will probably be given a copy of the state laws and guidelines governing the practice of professional engineering when you register. Even if the state board does not automatically supply you with a copy of the relevant laws, you should ask for one and read it thoroughly before embarking on your professional career. The law and guidelines clearly spell out the rights, responsibilities, and liabilities of the professional engineer.

As a registered professional engineer, you are entitled to be in responsible charge of engineering work in your field. You may supervise others in the execution of that work. You may offer and advertise your engineering services to the public, within the guidelines established by the profession.

In exercising your rights, your foremost responsibility is the protection of the public welfare in all aspects of your engineering work. These aspects include, but are not limited to, design, construction, advertising, relations with clients, and relations with other professionals. Each state has drawn up clear guidelines for conduct in all of these aspects of your work. The various professional engineering societies also have promulgated professional standards and ethical codes. You should be thoroughly familiar with existing guidelines, and keep current with any changes that may be made over time.

The profession of engineering is generally self-policing. Violations of professional ethical codes may be handled entirely within a professional society, for instance. However, violations of state regulations may call for disciplinary action by the state board. If it is determined that an individual is guilty of an alleged violation, possible actions by the state board include reprimand, suspension of the license for a specified period of time, and permanent revocation of the license. Revocation of a license may make it difficult or impossible for the individual to become licensed later in another state.

A professional engineer in responsible charge of design or other engineering work may be held legally liable for damages caused by product failure or other result of negligence. Liability insurance is strongly advised for all consulting engineers.

THE NEXT STEP

Your professional engineering license will open up a number of career options for you, from more rapid advancement within a manufacturing firm, to part-time consulting, to the establishment of your own consulting firm. Your choice among these options will obviously depend on your personal circumstances, goals, and preferences.

No matter which option you choose, it is very much to your advantage to join one or more of the professional engineering societies. The National Society of Professional Engineers (NSPE) represents the interests of all professional engineers in the United States, regardless of specialty. Every state has a professional engineering society (see Appendix C for the names and addresses of state societies). In addition, there are state, regional, and national societies for virtually every specialty in engineering. See Appendix B for the addresses of NSPE and some of the other national societies.

If you are considering establishing your own consulting practice, your professional society can help you with many aspects of the business, particularly advertising and referrals. However, getting established will be mostly up to you and to your own efforts.

There are many books on setting up and managing small businesses, which may be very helpful. In addition, there are a few books which focus directly on the issue of setting up a professional engineering firm. The American Consulting Engineers Council (ACEC) offers publications specifically oriented to the concerns of consulting engineers. In its series of books relevant to the professional engineer (which includes the booklet you are reading now), Professional Publications, Inc., offers *Getting Started as a Consulting Engineer*. With the help of the ideas to be found in these publications—and with your own energy, initiative, and engineering skill—you will be on your way to a successful practice as a professional engineer.

BIBLIOGRAPHY

Constance, John D. Engineers' Registration—Why and How. *Design News*, May 6, 1974, pp. 57 - 69.

Constance, John D. The Road to Registration 1: The Basic Requirements. *Machine Design*, September 4, 1975, pp. 54 - 59.

Constance, John D. The Road to Registration 2: Passing the Exam. *Machine Design*, September 18, 1975, pp. 82 - 85.

Constance, John D. Getting Licensed in Any State. *Chemical Engineering*, April 26, 1976, pp. 115 - 118.

Constance, John D. How Ready Are You for Registration? *Chemical Engineering*, May 24, 1976, pp. 153 - 158.

Constance, John D. How to Get Ready for Registration. *Chemical Engineering*, June 21, 1976, pp. 179 - 182.

Constance, John D. Documenting Your P.E. Qualifications. *Machine Design*, April 21, 1977, pp. 196-198.

Eckard, Joseph D., Jr., *Professional Engineer's License Guide: What You Need to Know and Do to Obtain PE (and EIT) Registration.* 3rd Edition. Boston, Massachusetts: Herman Publishing, Inc., 1978.

NCEE. *The Registration of Professional Engineers and Land Surveyors in the United States.* NCEE, 1978.

NCEE. *The Practice of Engineering in the United States.* NCEE, 1973.

NCEE. *Candidate Performance Summary.* Published yearly by NCEE.

NCEE. *The Fundamentals of Engineering Examination: Procedures Used by NCEE to Set a Minimum Passing Score.* Princeton, New Jersey: ETS, February 1981.

News from NCEE: Change in Setting Passing Scores to Raise Testing Standards. *Professional Engineers and Land Surveyors Report*, Spring 1984, p. 2.

Tow, Phil, What Is ABET? *Professional Engineers and Land Surveyors Report*, Spring 1984, p. 2.

APPENDIX A

STATE ENGINEERING REGISTRATION BOARDS

ALABAMA State Board of Registration for Professional Engineers and Land Surveyors
Executive Secretary
750 Washington Avenue, Suite 212
Montgomery, Alabama 36130
Tel: (205) 261-5568

ALASKA State Board of Registration for Architects, Engineers and Land Surveyors
Licensing Examiner
Pouch D
State Office Building, 9th Floor
Juneau, Alaska 99811
Tel: (907) 465-2540

ARIZONA State Board of Technical Registration
Executive Director
1645 W. Jefferson Street, Suite 140
Phoenix, Arizona 85007
Tel: (602) 255-4053

ARKANSAS State Board of Registration for Professional Engineers and Land Surveyors
Secretary-Treasurer
P.O. Box 2541
1818 W. Capitol
Little Rock, Arkansas 72203
Tel: (501) 371-2517

CALIFORNIA Board of Registration for Professional Engineers and Land Surveyors
Executive Officer
1428 Howe Street, Suite 56
Sacramento, California 95825
Tel: (916) 920-7466

COLORADO State Board of Registration for Professional Engineers and Professional Land Surveyors
Program Administrator
600-B State Services Building
1525 Sherman Street
Denver, Colorado 80203
Tel: (303) 866-2396

CONNECTICUT State Board of Examiners for Professional Engineers and Land Surveyors
Administrator
State Office Building
165 Capitol Avenue, Room G-3A
Hartford, Connecticut 06106
Tel: (203) 566-3386

DELAWARE Association of Professional Engineers
Executive Office Secretary
2005 Concord Pike
Wilmington, Delaware 19803
Tel: (302) 656-7311

DISTRICT OF COLUMBIA Board of Registration for Professional Engineers
Executive Secretary
614 H Street, N.W., Room 910
Washington, D.C. 20001
Tel: (202) 727-7454

FLORIDA State Board of Professional Engineers
Executive Director
130 North Monroe Street
Tallahassee, Florida 32301
Tel: (904) 488-9912

GEORGIA State Board of Registration for Professional Engineers and Land Surveyors
Executive Director
166 Pryor Street, S.W.
Atlanta, Georgia 30303
Tel: (404) 656-3926

GUAM Territorial Board of Registration for Professional Engineers, Architects and Land Surveyors
Chairman
Department of Public Works
Government of Guam, P.O. Box 2950
Agana, Guam 96910
Tel: (671) 646-8643

HAWAII State Board of Registration for Professional Engineers, Architects, Land Surveyors and Landscape Architects
Executive Secretary
Licensing Office
P.O. Box 3469
1010 Richards Street
Honolulu, Hawaii 96801
Tel: (808) 548-4100

IDAHO Board of Professional Engineers and Land Surveyors
Executive Secretary
842 La Cassia Drive
Boise, Idaho 83705
Tel: (208) 334-3860

ILLINOIS Department of Registration and Education
Division Manager
Professional Engineers' Examining Committee
320 W. Washington Street, 3rd Floor
Springfield, Illinois 62786
Tel: (217) 782-0177

INDIANA State Board of Registration for Professional Engineers and Land Surveyors
Executive Director
1021 State Office Building
100 North Senate Avenue
Indianapolis, Indiana 46204
Tel: (317) 232-1840

IOWA State Board of Engineering Examiners
Executive Secretary
1209 East Court Avenue
State Capital Complex
Des Moines, Iowa 50319
Tel: (515) 281-5602

KANSAS State Board of Technical Professions
Executive Secretary
214 West 6th Street, Second Floor
Topeka, Kansas 66603
Tel: (913) 296-3053

KENTUCKY State Board of Registration for Professional Engineers and Land Surveyors
Executive Director
Rt. 3-96
5 Millville Road
Frankfort, Kentucky 40601
Tel: (502) 564-2680 & (502) 564-2681

LOUISIANA State Board of Registration for Professional Engineers and Land Surveyors
Executive Secretary
1055 St. Charles Avenue, Suite 415
New Orleans, Louisiana 70130
Tel: (504) 568-8450

MAINE State Board of Registration for Professional Engineers
Secretary
State House
Augusta, Maine 04333
Tel: (207) 289-3236

MARYLAND State Board of Registration for Professional Engineers
Executive Secretary
501 St. Paul Place, Room 902
Baltimore, Maryland 21202
Tel: (301) 659-6322

MASSACHUSETTS State Board of Registration of Professional Engineers and Land Surveyors
Office Secretary
Leverett Saltonstall Building, Room 1512
100 Cambridge Street
Boston, Massachusetts 02202
Tel: (617) 727-3055

MICHIGAN Board of Professional Engineers
Administrative Secretary
P.O. Box 30018
611 West Ottawa
Lansing, Michigan 48909
Tel: (517) 373-3880

MINNESOTA State Board of Registration for Architects, Engineers, Land Surveyors and Landscape Architects
Executive Secretary
5th Floor, Metro Square
St. Paul, Minnesota 55101
Tel: (612) 296-2388

MISSISSIPPI State Board of Registration for Professional Engineers and Land Surveyors
Office Manager
P.O. Box 3
200 South President Street, Suite 516
Jackson, Mississippi 39205
Tel: (601) 354-7241

MISSOURI Board of Architects, Professional Engineers and Land Surveyors Secretary-Treasurer
P.O. Box 184
3523 North Ten Mile Drive
Jefferson City, Missouri 65102
Tel: (314) 751-2334

MONTANA State Board of Professional Engineers and Land Surveyors
Administrative Secretary
Department of Commerce
1424-9th Avenue
Helena, Montana 59620-6521
Tel: (406) 444-4285

NEBRASKA State Board of Examiners for Professional Engineers and Architects
Executive Director
P.O. Box 94751
301 Centennial Mall, South
Lincoln, Nebraska 68509
Tel: (402) 471-2021 & (402) 471-2407

NEVADA State Board of Registered Professional Engineers and Land Surveyors
Executive Secretary
1755 East Plumb Lane, Suite 102
Reno, Nevada 89502
Tel: (702) 329-1955

NEW HAMPSHIRE State Board of Professional Engineers
Secretary
Storrs Street
Concord, New Hampshire 03301
Tel: (603) 271-2219

NEW JERSEY State Board of Professional Engineers and Land Surveyors
Executive Secretary
1100 Raymond Boulevard
Newark, New Jersey 07102
Tel: (201) 648-2660

NEW MEXICO State Board of Registration for Professional Engineers and Land Surveyors
Administrator-Secretary
P.O. Box 4847
Santa Fe, New Mexico 87502
Tel: (505) 827-9940

NEW YORK State Board for Engineering and Land Surveying
Executive Secretary
The State Education Department
Cultural Education Center
Madison Avenue
Albany, New York 12230
Tel: (518) 474-3846

NORTH CAROLINA Board of Registration for Professional Engineers and Land Surveyors
Executive Secretary
3620 Six Forks Road
Raleigh, North Carolina 27609
Tel: (919) 781-9499

NORTH DAKOTA State Board of Registration for Professional Engineers and Land Surveyors
Executive Secretary
P.O. Box 1357
420 Avenue B East
Bismark, North Dakota 58501
Tel: (701) 258-0786

OHIO State Board of Registration for Professional Engineers and Surveyors
Executive Secretary
65 South Front Street, Room 302
Columbus, Ohio 43215
Tel: (614) 466-8948

OKLAHOMA State Board of Registration for Professional Engineers and Land Surveyors
Executive Secretary
Oklahoma Engineering Center, Room 120
201 N.E. 27th Street
Oklahoma City, Oklahoma 73105
Tel: (405) 521-2874

OREGON State Board of Engineering Examiners
Executive Secretary
Department of Commerce, Room 403
Labor and Industries Building
Salem, Oregon 97310
Tel: (503) 378-4180

PENNSYLVANIA State Registration Board for Professional Engineers
Administrative Secretary
P.O. Box 2649
Transportation & Safety Building, 6th Floor
Commonwealth Avenue & Forester Street
Harrisburg, Pennsylvania 17105-2649
Tel: (717) 783-7049

PUERTO RICO Board of Examiners of Engineers, Architects, and Surveyors
Administrative Officer
Box 3271
Tanca Street, Comer Tetuan
San Juan, Puerto Rico 00904
Tel: (809) 722-2121

RHODE ISLAND State Board of Registration for Professional Engineers and Land Surveyors
Administrative Assistant
308 State Office Building
Providence, Rhode Island 02903
Tel: (401) 277-2565

SOUTH CAROLINA State Board of Engineering Examiners
Agency Director
2221 Devine Street, Suite 404
P.O. Drawer 50408
Columbia, South Carolina 29205
Tel: (803) 758-2855

SOUTH DAKOTA State Commission of Engineering and Architectural Examiners
Executive Secretary
2040 West Main Street, Suite 212
Rapid City, South Dakota 57701
Tel: (605) 394-2510

TENNESSEE State Board of Architectural and Engineering Examiners
Executive Assistant
546 Doctors' Building
706 Church Street
Nashville, Tennessee 37219-5322
Tel: (615) 741-3221

TEXAS State Board of Registration for Professional Engineers
Executive Director
1917 IH 35 South
P.O. Drawer 18329
Austin, Texas 78760
Tel: (512) 475-3141

UTAH State Board of Registration for Professional Engineers and Land Surveyors
Director
Division of Registration
P.O. Box 45802
160 East 300 South
Salt Lake City, Utah 84145
Tel: (801) 530-6632

VERMONT State Board of Registration for Professional Engineers
Division of Licensing and Regulations
Pavilion Building
Montpelier, Vermont 05602
Tel: (802) 828-2363

VIRGINIA State Board of Architects, Professional Engineers, Land Surveyors and Certified Landscape Architects
Assistant Director
Department of Commerce
Seaboard Building, 5th Floor
3600 West Broad Street
Richmond, Virginia 23230-4917
Tel: (804) 257-8512

VIRGIN ISLANDS Board for Architects, Engineers and Land Surveyors
Secretary
Submarine Base, P.O. Box 476
St. Thomas, Virgin Islands 00801
Tel: (809) 774-1301

WASHINGTON State Board of Registration for Professional Engineers and Land Surveyors
Executive Secretary
P.O. Box 9649
9th & Columbia Building, 3rd Floor
Olympia, Washington 98504
Tel: (206) 753-6966

WEST VIRGINIA State Board of Registration for Professional Engineers
Executive Director
608 Union Building
Charleston, West Virginia 25301
Tel: (304) 348-3554

WISCONSIN State Examining Board of Architects, Professional Engineers, Designers, and Land Surveyors
P.O. Box 8936
1400 East Washington Avenue
Madison, Wisconsin 53708
Tel: (608) 266-1397

WYOMING State Board of Examining Engineers
Secretary-Accountant
Barrett Building
Cheyenne, Wyoming 82002
Tel: (307) 777-6156

APPENDIX B

A PARTIAL LIST OF NATIONAL ENGINEERING SOCIETIES AND OTHER ENGINEERING-RELATED ORGANIZATIONS

AACE: American Association of Cost Engineers. 308 Monongahela Building, Morgantown, West Virginia 26505.

AAES: American Association of Engineering Societies. A private, non-profit organization to which all of the major engineering societies belong. AAES advises, represents, and fosters communication among the societies. 345 East 47th Street, New York, New York 10017. Telephone: (212) 705-7840.

ABET: Accreditation Board for Engineering and Technology. A non-profit, independent organization which is the primary accrediting organization of engineering degree programs. 345 East 47th Street, New York, New York 10017-2397.

ACEC: American Consulting Engineers Council. 1015 15th Street, N.W., Washington, D.C. 20005. Telephone: (202) 347-7474.

AIAA: American Institute of Aeronautics and Astronautics. 1633 Broadway, New York, New York 10019. Telephone: (212) 581-4300.

AIChE: American Institute of Chemical Engineers. 345 East 47th Street, 12th Floor, New York, New York 10017. Telephone: (212) 705-7338.

AIMMPE: American Institute of Mining, Metallurgical, and Petroleum Engineers. 345 East 47th Street, New York, New York 10017. Telephone: (212) 705-7695.

AIPE: American Institute of Plant Engineers. 3975 Erie Avenue, Cincinnati, Ohio 45208. Telephone: (513) 561-6000.

ANS: American Nuclear Society. 555 N. Kensington, LaGrange Park, Illinois 60525. Telephone: (312) 352-6611.

APCA: Air Pollution Control Association. P.O. Box 2861, Pittsburgh, Pennsylvania 15213. Telephone: (412) 232-3444.

ASAE: American Society of Agricultural Engineers. 2950 Niles Road, St. Joseph, Michigan 49085. Telephone: (616) 429-0300.

ASCE: American Society of Civil Engineers. 345 East 47th Street, New York, New York 10017. Telephone: (212) 705-7496.

ASCET: American Society of Certified Engineering Technicians. P.O. Box 7789, Shawnee Mission, Kansas 66206. Telephone: (913) 451-4938.

ASEE: American Society for Engineering Education. Eleven DuPont Circle, Suite 200, Washington, D.C. 20036. Telephone: (202) 293-7080.

ASHRAE: American Society of Heating, Refrigerating, and Air-Conditioning Engineers. 1791 Tullie Circle, N.E., Atlanta, Georgia 30329. Telephone: (404) 636-8400.

ASME: American Society of Mechanical Engineers. 345 East 47th Street, New York, New York 10017. Telephone: (212) 705-7722.

ASNT: American Society of Non-destructive Testing. 4153 Arlingate Plaza, Caller #28518, Columbus, Ohio 43228-0518. Telephone: (614) 274-6003.

ASPE: American Society of Plumbing Engineers. 15233 Ventura Boulevard, Sherman Oaks, California 91403. Telephone: (818) 783-4845.

ASQC: American Society for Quality Control. 230 Wells Street, Milwaukee, Wisconsin 53203. Telephone: (414) 272-8575.

ASSE: American Society of Safety Engineers. 850 Busse Highway, Park Ridge, Illinois 60068. Telephone: (312) 692-4121.

AWS: American Welding Society. 550 N.W. LeJeune Road, Miami, Florida 33126. Telephone: (305) 443-9353.

ETCI: Engineering Technologist Certification Institute. An independent, non-profit certifying organization sponsored by NSPE which examines engineering technologists and awards certificates of competence. 1420 King Street, Alexandria, Virginia. Telephone: (703) 604-2835.

IEEE: Institute of Electrical and Electronic Engineers. 345 East 47th Street, New York, New York 10017. Telephone: (212) 705-7900.

IIE: Institute of Industrial Engineers. 25 Technology Park/Atlanta, Norcross, Georgia 30092. Telephone: (404) 449-0460.

NCEE: National Council of Engineering Examiners. The organization which writes and distributes the uniform EIT and PE examinations. P.O. Box 1686, Clemson, South Carolina 29633. Telephone: (803) 654-6824.

NICET: National Institute for the Certification of Engineering Technicians. An independent, non-profit certifying organization sponsored by NSPE which examines engineering technicians and awards certificates of competence. 1420 King Street, Alexandria, Virginia. Telephone: (703) 684-2835.

NSPE: National Society of Professional Engineers. A national professional society encompassing all state PE societies. NSPE represents the social, economic, and political interests of all engineers in the United States. 1420 King Street, Alexandria, Virginia 22314. Telephone: (703) 684-2800.

Professional Engineering Institute: A non-profit educational organization which provides assistance, study materials, and state-approved review courses to engineers seeking registration. P.O. Box 639, San Carlos, California 94070.

SFPE: Society of Fire Protection Engineers. 60 Batterymarch Street, Boston, Massachusetts 02110.

SME: Society of Manufacturing Engineers. P.O. Box 930, Dearborn, Michigan 48121. Telephone: (313) 271-1500.

SPHE: Society of Packaging and Handling Engineers. 11800 Sunrise Valley Drive, Reston International Center, Suite 212, Reston, Virginia 22091. Telephone: (703) 620-9380.

APPENDIX C
ADDRESSES OF STATE PROFESSIONAL ENGINEERING SOCIETIES

ALABAMA

Alabama Society of Professional Engineers
Business/Engineering Building, Suite 255 ERG
1000 Eleventh Way, South
Birmingham, Alabama 35294
Telephone: (205) 934-8470

ALASKA

Alaska Society of Professional Engineers
c/o University of Alaska
Duckering Building, Room 1331
306 Tanana Drive
Fairbanks, Alaska 99701
Telephone: (907) 349-6572 (Anchorage)
(907) 474-7330 (Fairbanks)

ARIZONA

Arizona Society of Professional Engineers
2415 West Colter #4
Phoenix, Arizona 85015
Telephone: (602) 249-0963

ARKANSAS

Arkansas Society of Professional Engineers
One Union Station, Suite 409
Markham & Victory Streets
Little Rock, Arkansas 72201
Telephone: (501) 376-4128

CALIFORNIA

California Society of Professional Engineers
1005-12th Street, Suite H
Sacramento, California 95814
Telephone: (916) 442-1041

COLORADO

Professional Engineers of Colorado
2701 Alcott Street, Suite 263
Denver, Colorado 80211
Telephone: (303) 458-0465

CONNECTICUT

Connecticut Society of Professional Engineers
2600 Dixwell Avenue
Hamden, Connecticut 06514
Telephone: (203) 281-4322

DELAWARE

Delaware Society of Professional Engineers
Post Office Box 2865, Suite 215
Wilmington, Delaware 19805
Telephone: (302) 656-7311

DISTRICT OF COLUMBIA

District of Columbia Society of Professional Engineers
Post Office Box 60703
Washington, D.C. 20039-0703
Telephone: (301) 946-7330

FLORIDA

Florida Engineering Society
125 South Gadsden Street
Post Office Box 750
Tallahassee, Florida 32302
Telephone: (904) 224-7121

GEORGIA

Georgia Society of Professional Engineers
Two Park Place, Suite 234
1888 Emery Street, N.W.
Atlanta, Georgia 30318
Telephone: (404) 355-0177

GUAM

Guam Society of Professional Engineers
Post Office Box 20670
GMF Guam 96921
Telephone: 011-(617) 646-4524

HAWAII

Hawaii Society of Professional Engineers
Post Office Box 2750
Honolulu, Hawaii 96840
Telephone: (808) 548-3570

IDAHO

Idaho Society of Professional Engineers
842 LaCassia Drive
Boise, Idaho 83705
Telephone: (208) 345-1730

ILLINOIS

Illinois Society of Professional Engineers
1304 South Lowell
Springfield, Illinois 62704
Telephone: (217) 544-7424

INDIANA

Indiana Society of Professional Engineers
225 East North Street, Suite A
Indianapolis, Indiana 46204
Telephone: (217) 544-7424

IOWA

Iowa Engineering Society
1051 Office Park Road, Suite 2
West Des Moines, Iowa 50265
Telephone: (515) 223-0309

KANSAS

Kansas Engineering Society
Post Office Box 477
Topeka, Kansas 66601
Telephone: (913) 233-1867

KENTUCKY

Kentucky Society of Professional Engineers
Post Office Box 458
Frankfort, Kentucky 40602
Telephone: (502) 695-5680

LOUISIANIA

Louisiana Engineering Society
Post Office Box 2683
Baton Rouge, Louisiana 70821
Telephone: (504) 344-4318

MAINE

Maine Society of Professional Engineers
RR #2, Box 5760
Oxford, Maine 04270
Telephone: (207) 998-2730

MARYLAND

Maryland Society of Professional Engineers
908 Rappaix Court
Towson, Maryland 21204
Telephone: (301) 828-0720

MASSACHUSETTS

Massachusetts Society of Professional Engineers, Inc.
555 Huntington Avenue
Boston, Massachusetts 02115
Telephone: (617) 442-7745

MICHIGAN

Michigan Society of Professional Engineers
Post Office Box 10214
Lansing, Michigan 48933
Telephone: (517) 487-9388

MINNESOTA

Minnesota Society of Professional Engineers
555 Park Street, Suite 130
St. Paul, Minnesota 55103
Telephone: (612) 292-8860

MISSISSIPPI

Mississippi Engineering Society
5425 Executive Place, Suite D
Jackson, Mississippi 39206
Telephone: (601) 366-1312

MISSOURI

Missouri Society of Professional Engineers
330 East High Street
Jefferson City, Missouri 65101
Telephone: (314) 636-4861

MONTANA

Montana Society of Engineers
1629 Avenue D
Post Office Box 20996
Billings, Montana 59104
Telephone: (406) 259-7300

NEBRASKA

Nebraska Society of Professional Engineers
1630 K Street, Suite D
Lincoln, Nebraska 68508
Telephone: (402) 476-2572

NEVADA

Nevada Society of Professional Engineers
3593 Tioga Way
Las Vegas, Nevada 89109
Telephone: (702) 735-0003

NEW HAMPSHIRE

New Hampshire Society of Professional Engineers
93 Belknap Street
Dover, New Hampshire 03820
Telephone: (603) 668-8223

NEW JERSEY

New Jersey Society of Professional Engineers
226 West State Street
Trenton, New Jersey 08608
Telephone: (609) 393-0099

NEW MEXICO

New Mexico Society of Professional Engineers
1615 University Boulevard, N.E.
Albuquerque, New Mexico 87102
Telephone: (505) 247-9181

NEW YORK

New York State Society of Professional Engineers
150 State Street 3rd Floor
Albany, New York 12207
Telephone: (518) 465-7386

NORTH CAROLINA

Professional Engineers of North Carolina
4000 Wake Forest Road, Suite 116
Raleigh, North Carolina 27609
Telephone: (919) 872-0683

NORTH DAKOTA

North Dakota Society of Professional Engineers
Post Office Box 1031
Grand Forks, North Dakota 58206-1031
Telephone: (701) 777-3782

OHIO

Ohio Society of Professional Engineers
445 King Avenue, Room 103
Columbus, Ohio 43201
Telephone: (614) 424-6640

OKLAHOMA

Oklahoma Society of Professional Engineers
Oklahoma Engineering Center
201 N.E. 27th Street, Room 125
Oklahoma City, Oklahoma 73105
Telephone: (405) 528-1435

OREGON

Professional Engineers of Oregon
1423 S.W. Columbia
Portland, Oregon 97201
Telephone: (503) 228-2701

PANAMA CANAL ZONE

Panama Canal Society of Professional Engineers
P.O. Box 6-4455, El Dorado
Panama, Republic of Panama
Telephone: (507) 56-6681

PENNSYLVANIA

Pennsylvania Society of Professional Engineers
4303 Derry Street
Harrisburg, Pennsylvania 17111
Telephone: (717) 561-0590

PUERTO RICO

Puerto Rico Society of Professional Engineers
1798 Astromelia
URB. Mans. De Rio Piedras
Rio Piedras, Puerto Rico 00926
Telephone: (809) 761-6014

RHODE ISLAND

Rhode Island Society of Professional Engineers
10 Orms Street
Providence, Rhode Island 02904
Telephone: (401) 751-3200

SOUTH CAROLINA

South Carolina Society of Professional Engineers
Post Office Box 11937
Columbia, South Carolina 29211
Telephone: (803) 771-4271

SOUTH DAKOTA

South Dakota Engineering Society
110 West Capitol, Box 1037
Pierre, South Dakota 57501
Telephone: (605) 224-1591

TENNESSEE

Tennessee Society of Professional Engineers
206 Capitol Boulevard
Suite #301
Nashville, Tennessee 37219
Telephone: (615) 242-2486

TEXAS

Texas Society of Professional Engineers
Post Office Box 2145
Austin, Texas 78768
Telephone: (512) 472-9286

UTAH

Utah Society of Professional Engineers
P.O. Box 27434
Salt Lake City, Utah 84127-0434
Telephone: (801) 535-4144

VERMONT

Vermont Society of Professional Engineers
RFD #1, Box 489
Williamstown, Vermont 05679
Telephone: (802) 728-3376

VIRGINIA

Virginia Society of Professional Engineers
116-118 East Franklin Street, Suite 601
Richmond, Virginia 23219
Telephone: (804) 780-2491

WASHINGTON

Washington Society of Professional Engineers
12828 Northup Way
Bellevue, Washington 98005
Telephone: (206) 885-2660

WEST VIRGINIA

West Virginia Society of Professional Engineers
301 Adams Street, Room 508
Fairmont, West Virginia 26554
Telephone: (304) 366-9785

WISCONSIN

Wisconsin Society of Professional Engineers
1045 East Dayton Street, Room 104
Madison, Wisconsin 53703
Telephone: (608) 251-7872

WYOMING

Wyoming Society of Professional Engineers
1915 Inverness Boulevard
Rawlins, Wyoming 82301
Telephone: (307) 266-4346

APPENDIX D

GLOSSARY OF PROFESSIONAL ENGINEERING TERMS

Accredited Engineering Program. A program of studies in engineering which is accredited by ABET.

Board: See 'Board of Registration'.

Board of Registration: An agency within each state government which administers the EIT and PE exams, maintains a roster of registered engineers, investigates complaints against engineers, and disciplines engineers who violate state regulations governing the practice of engineering.

Branch: See 'Discipline'.

C.E.: Abbreviation for 'Consulting Engineer' and 'Civil Engineer'.

Certification: The process of obtaining a certificate of competence from an organization, usually a professional society. Certification has no legal basis and is not a substitute for registration.

CET: Abbreviation for 'Certified Engineering Technician'.

Chartered Engineers' Act: A registration act similar to those of Delaware and Canada, characterized by self-regulation of the engineering profession.

Comity: See 'Reciprocity'.

Consulting Engineer: An engineer who provides services for a fee without becoming an employee.

Discipline: A branch of engineering recognized by a state for purposes of examination or licensing.

E-I-T or EIT: Abbreviation for 'Engineer-in-Training'.

Engineer-in-Training Examination: The first of two 8-hour examinations which most engineers must take to obtain a PE license. The EIT examination covers all technical subjects normally found in the first three years of an undergraduate (B.S.) degree program in engineering. Some individuals possessing the proper combined qualifications of age, experience, and education may skip the EIT exam.

Expert Witness: A technical specialist who testifies in a court of law on behalf of a client or as a friend of the court. Many states require expert witnesses who testify on engineering matters to be registered as professional engineers.

FE Exam: See 'Fundamentals of Engineering Examination'.

Fundamentals of Engineering Examination: The name given by NCEE to the EIT examination.

Grandfather Clause: A clause in legislation defining new engineering disciplines which allows registration of qualified applicants for a short time without registration examinations.

Grandfathering: The act of obtaining a PE license in a new engineering discipline without taking an examination.

I.E. Examination: See 'Intern Engineer Examination'.

Individual Exam: A shortened examination administered by the state board to engineers with substantial age and experience. The individual exam usually consists of an oral interview and a short written test.

Industrial Exemption: A regulation in most states which exempts engineers employed in industry from the requirement of registration.

Intern Engineer Examination: The name given by some states to the EIT exam.

Licensing: See 'Registration'.

Manufacturers' Exemption: See 'Industrial Exemption'.

Model Act: An engineers' act proposed by NCEE for adoption by all states. The model act provides for single-title registration and the elimination of the industrial exemption. Thus far, only a few states have adopted the provisions of the model act.

Model Law: See 'Model Act'.

National Engineering Certificate: A certificate issued by NCEE to registered engineers which documents verified education and work experience.

NEC: See 'National Engineering Certificate'.

P.E.: A protected abbreviation for 'Professional Engineer'.

P.E. Examination: An 8-hour written examination which is taken after the EIT examination (or after the EIT examination is waived) to determine who may use the title 'Professional Engineer'.

Practice Act: A state regulation which prohibits unregistered engineers from working as consultants in specified fields.

Practice Protection: The prohibition against working as a consulting engineer without being registered.

Principles and Practice Examination: The name given by NCEE to the PE examination.

The Public: Any person or group of people which makes use of services regulated by a state's Department of Occupational Licensing or similar department.

Public Member: A member of the Board of Registration who is not an engineer.

R.E.: A non-protected abbreviation for 'Registered Engineer'. Many engineers refer to the P.E. exam as the 'R.E. Exam'.

Reciprocal License: A professional engineering license granted without examination to an engineer already registered in another state.

Reciprocity: The act of granting a reciprocal license without examination to an engineer already registered in another state.

Registration: The process of obtaining a professional engineering license and specific legal rights from the Board of Registration. Registration always results in title protection and it may result in practice protection.

Registration by Discipline: A registration process whereby an engineer receives a title in a specific discipline (e.g., 'Professional Electrical Engineer') and is legally limited to working in that discipline.

Responsible Charge of Work: The independent control and direction of professional engineering work.

Seal: See 'Stamp'.

Single-Title Registration: A registration process in which all registered engineers receive the title 'Professional Engineer' regardless of discipline. Engineers are ethically bound not to practice outside their fields of expertise.

Stamp: The imprint from a rubber stamp or similar device containing an engineer's name, discipline, and registration number. The format of the stamp is specified by the Board of Registration.

State Board: See 'Board of Registration'.

Title Act: A regulation which prohibits a specific title from being used by unregistered engineers.

Title Protection: Protection given to specific titles by prohibiting their use by unregistered engineers.

Uniform Examination: An examination (EIT or PE) developed and distributed by NCEE and used by almost all states.